Praise for

Panic in Level 4

"Darkly fascinating . . . [*Panic in Level 4*] is sure to please science fanatics, or anyone else obsessed with nature's murkier mysteries."
—Newsweek.com

"With his 1994 sensation *The Hot Zone,* science writer Richard Preston terrified millions by describing how the Ebola virus left victims hemorrhaging from every orifice. In [*Panic in Level 4*], he continues to probe nature's stranger side, including what DNA tastes like and a disease that compels people to chew their own flesh."
—*USA Today*

"[Richard Preston's] stories sparkle with images of stark beauty and darkness; mature reflections about the complex worlds we all occupy. Without being aware of it . . . you find yourself reconsidering your own previously held assumptions on matters of great consequence: our humanity, the tyranny of illness, the complexity of nature, the elusive search for meaning, the compulsions and fixations that grab hold of us, and the miracle of human intellect. [Preston's] scientific writing feels like a reli-

gious journey of sorts, not really a search for God, but rather an appreciation of what seems almost God-like within each man."
—*The Denver Post*

"[Preston's] profiles of the stories' human subjects probe deeply. . . . A fast and entertaining read." —*The Charlotte Observer*

"Why scare ourselves with ghosts or vampires when we can really get freaked out with parasitic insects and Ebola? [Richard Preston] probes curiosities like the "self-cannibals," who . . . are compelled to chew on their own flesh due to a DNA anomaly, and two computer-whiz brothers who built a monster computer in their apartment." —*The Boston Globe*

"'I love exploring unseen worlds,' Preston has said. That inclination combined with his sure-handed writing touch has put him in the pantheon of creative nonfiction authors, including Gay Talese, Tracy Kidder and John McPhee. . . . By illuminating the dark edges of our world, Preston puts us in touch with our humanity." —*The Free Lance–Star* (Fredericksburg, VA)

"The *Hot Zone* author does what he does so well in this new nonfiction collection, writing about bizarre illnesses and plagues and scaring the rest of us silly." —*The Seattle Times*

"Hard to put down . . . [*Panic in Level 4*] takes [Preston's *New Yorker* offerings] to new levels of interest."
—*Wisconsin State Journal*

"Preston personifies perceptiveness and empathy in journalism."
—*Booklist*

"Preston's fans will enjoy his showing how few degrees of separation there are between far-flung areas of scientific endeavors."
—*Publishers Weekly*

"Well researched, well paced and accessible." —*Kirkus Reviews*

ALSO BY RICHARD PRESTON

First Light
American Steel
The Hot Zone
The Cobra Event
The Demon in the Freezer
The Boat of Dreams: A Christmas Story
The Wild Trees

Panic in Level 4

PANIC IN LEVEL 4

Cannibals, Killer Viruses, and Other Journeys to the Edge of Science

Richard Preston

RANDOM HOUSE
TRADE PAPERBACKS
NEW YORK

2009 Random House Trade Paperback Edition

Published in the United States by Random House Trade Paperbacks, an imprint of The Random House Publishing Group, a division of Random House, Inc., New York.

RANDOM HOUSE TRADE PAPERBACKS and colophon are trademarks of Random House, Inc.

Originally published in hardcover in the United States by Random House, an imprint of The Random House Publishing Group, a division of Random House, Inc., in 2008.

Portions of this book appeared in different form in *The New Yorker.*

Grateful acknowledgment is made to Jean-François Ruppol, M.D., for permission to reprint an excerpt from his unpublished narrative "Ebola 2," translated into English from French by Richard Preston and William T. Close. Used by permission of Jean-François Ruppol, M.D.

LIBRARY OF CONGRESS CATALOGING-IN-PUBLICATION DATA

Preston, Richard
Panic in level 4/Richard Preston.
p. cm.
"Portions of this book appeared in different form in *The New Yorker.*"
ISBN 978-0-8129-7560-4
1. Medicine, Popular. 2. Science. 3. Science writers. I. Title.
RC81.P856 2008 616.02'4—dc22 2007041770

Printed in the United States of America

www.atrandom.com

987654321

To my father,
Jerome Preston, Jr.

Considering the marvelous adventures
you took me on and sent me on,
this book seems inevitable.

Contents

List of Illustrations

Introduction: Adventures in Nonfiction Writing

OLIVER HEAVISIDE, the English mathematician and physicist, once said, "In order to know soup, it is not necessary to climb into a pot and be boiled." Unfortunately, this statement is not true for journalists. As a writer of what's called "literary nonfiction" or "creative nonfiction"—narrative that is said to read like a novel but is factually verifiable—it has often been my practice to climb into the soup. Getting boiled with your characters is a good way to get to know them, but it has occasionally led me into frightening situations.

Some years ago, while I was researching *The Hot Zone,* a book that focuses on the Ebola virus, I may have had a meeting with an unknown strain of Ebola. (A virus is an exceedingly small life-form, an infectious parasite that replicates inside living cells, using the cell's own machinery to make more copies of itself.)

Ebola has now been classified into seven different known types. Though it has been studied for more than thirty years, Ebola is one of the least-understood viruses in nature. Scientists have been understandably reluctant to study Ebola too closely because it has on occasion killed those who tried to do so. The virus was first was noticed in 1976, when it surfaced in Yambuku, Zaire (now the Democratic Republic of Congo), near the Ebola River, where it sacked a Catholic mission hospital, killing most of the medical staff along with a number of patients and people the patients came into contact with. Ebola spreads from one person to the next by direct contact with blood or secretions, including sweat. There is no evidence that it can spread among humans through the air, although there is some evidence that it may spread among monkeys this way. As a parasite, Ebola carries on its life cycle in some unidentified type of animal—Ebola's natural host—that lives in certain unidentified habitats in equatorial Africa. Occasionally Ebola comes into contact with a person, and the virus makes what is known as a trans-species jump from its host into the human species.

When Ebola gets inside a human host, it causes the person's immune system to vanish, and the person dies with hemorrhages coming from the body's orifices. The most lethal strains of Ebola have been known to kill up to 95 percent of people who become infected with it. Ebola causes people to vomit masses of black blood with a distinctive "coffee grounds" appearance. Victims can have a bright red nosebleed, or epistaxis; it won't stop. A spotty, bumpy rash spreads over the body, while small, starlike hemorrhages appear beneath the skin. An Ebola patient can have blood standing in droplets on the eyelids and running from the tear ducts down the face. Blood can flow from the nose, mouth, vagina, rectum. The testicles can become infected with Ebola and can swell up or be destroyed. Victims display signs of psychosis. They can develop endless hiccups. Rarely, in particularly severe cases of Ebola, the linings of the intestines and rectum may come off. Those membranes may be expelled through the anus in raglike pieces called casts, or the intestinal lining can emerge in the form of a sleeve, like a sock. When an Ebola patient expels a sleeve, it is known as throwing a tubular cast.

SOME OF THE ACTION in *The Hot Zone* takes place at Fort Detrick, an Army base in the rolling country along the eastern flank of the Appalachian Mountains in Maryland, an hour's drive northwest of Washington, D.C. The Army's Level 4 virus laboratories at Fort Detrick are clustered inside a large, nearly windowless building that sits near the eastern perimeter of the base. This building is the headquarters of the United States Army Medical Research Institute of Infectious Diseases, or USAMRIID—a facility that Army people often simply refer to as "the Institute." While I was visiting USAMRIID, or the Institute, to interview various experts in Ebola, I began asking officials at the base for permission to put on a biohazard space suit and enter one of the Army's Biosafety Level 4 virus laboratories. I wanted to get a firsthand look at researchers handling Ebola. More than that, I wanted to know what it feels like to wear a biohazard space suit and be face-to-face with a real Level 4 virus, so that I could get a sense of my characters' feelings and experiences, and could describe them with convincing precision. (Many nonfiction writers refer to their characters as "subjects," but I prefer to think of them as dramatis personae in a true story.)

Biosafety Level 4, also called BL-4 or Level 4, is the highest and tightest level of biosecurity in a laboratory. Laboratories rated at Biosafety Level 4 are the repositories of viruses called hot agents—lethal viruses for which there is no vaccine or effective cure. Level 4 labs are sealed off from the outside world. People who go inside a Level 4 lab are required to wear a biohazard space suit, a pressurized whole-body suit, like an astronaut's, made of soft, flexible plastic, typically blue. Army researchers sometimes refer to the space suit as a "blue suit" because of its color. A soft, flexible helmet that completely surrounds the head is joined to the suit, and the helmet has a clear, flexible plastic faceplate in it; the suit also has an independent air supply. The air supply prevents you from breathing the air inside the lab, which could be contaminated with a hot agent. Ebola virus is classified as a Level 4 hot agent, one of the most dangerous known, and it has the potential to be used as a biological weapon. The scientists at USAMRIID (it's pronounced "you-SAM-rid") conduct research into vaccines and drugs that could be used to protect the population of the United States against a terrorist or military attack with a biological weapon, including Ebola virus. This is medical research, peaceful in nature.

The Level 4 labs inside the Institute consist of groups of interconnected rooms. Each group of rooms is known as a hot suite or a hot zone. Each suite is sealed off from the outside world and is accessible only through an air lock. The air lock has heavy, stainless steel doors. Inside the air lock there is a chemical decontamination shower, also known as a "decon" shower. The purpose of the decon shower is to sterilize the outer surface of the space suits of researchers who are leaving a hot zone, to prevent a hot agent from getting a ride to the outside world. Chest freezers inside the Army's hot zones are filled with collections of microvials (tiny plastic test tubes the size of a pencil stub) that contain frozen or freeze-dried samples of many different strains of lethal viruses.

The freezers are kept at 95 degrees below zero Fahrenheit. They are said to be hot. They are hot in a biological sense: they contain frozen samples of lethal viruses that are held in suspended animation in the extreme cold. The virus collections stored in USAMRIID's hot freezers are said to include strains of Bolivian hemorrhagic fever virus, Guanarito virus, Junín virus, dengue hemorrhagic fever, Venezuelan equine encephalitis (VEE), Japanese encephalitis virus (JEV), Hendra virus, Nipah virus, Lassa virus, and the seven known types of Ebola. (These viruses' effects on humans vary. Guanarito, Junín, and Lassa, for example, cause hemorrhaging from the body's orifices, like Ebola. VEE and JEV infect the brain and spinal cord, causing coma or death. Nipah is a brain virus from Malaysia that can trigger a literal meltdown of the brain. The brain of a Nipah victim can be semiliquefied as the virus consumes it, and can pour out of the skull during an autopsy.) The hot freezers also may contain (although the Army doesn't say much about this) an assortment of Level 4 Unknown viruses, or X viruses. The Unknown X viruses *appear* to be lethal in humans, but little is known about them. They've never been fully studied or classified. They may include something known as the Linköping Samples, which may or may not harbor an unidentified type of Ebola. The X viruses are presumed to be potentially lethal, so they are kept in Level 4 for safety.

We know that Ebola virus was one of the more powerful bioweapons in the arsenal of the old Soviet Union. In the years before the Soviet Union broke up, in 1991, bioweaponeers had reportedly been experimenting with aerosol Ebola—powdered, weaponized Ebola that could be dispersed through the air, over a city, for example. The Soviet weaponized Ebola was apparently stable enough that it could drift for distances in the air and still infect people through the lungs when they breathed a few particles of it. This is why the U.S. Army was studying it: the Army researchers were trying to come up with a vaccine or a drug treatment for Ebola, in case of a terrorist or military attack on the United States with Ebola.

As a natural disease, Ebola virus does not seem to be able to pass from person to person through the air, though. In each of the natural Ebola outbreaks, the disease seems to burn itself out, and Ebola fades away and is lost in the backdrop of nature, until, once again, by chance, it finds its way into a human host.

ONE DAY I WAS INTERVIEWING the commander of Fort Detrick. We were getting toward the end of the interview, and I decided to ask one last question. "I'd like to try to convey to readers what it really feels like to be face-to-face with Ebola virus." I said. "Could I go into Level 4?"

"That should be no problem," the commander answered promptly. "We'll get you outfitted in a blue suit," he said, "and walk you through Suite AA-4"—one of the Ebola hot zones. "It's down and cold," he added.

"What do you mean by 'down and cold'?" I asked.

He explained that the hot zone had been completely sterilized with gas and opened up for routine maintenance. The rooms weren't dangerous. Anyone could go into the lab without wearing a space suit. The hot freezers, too, had been moved out of the lab. Therefore, the lab was completely cold and safe.

"That's not really what I had in mind," I said.

"What did you have in mind?" he asked.

"I would like to experience the real thing, so that I can describe it. I'd like to go into a hot BL-4 lab and see how the scientists work with real Ebola."

"That's not possible," he answered immediately.

Security at USAMRIID was extremely tight. Even so, it was not as tight as it would become. That day in the commander's office was some nine years before the anthrax terror attacks of the autumn of 2001, shortly after 9/11. The anthrax attacks came to be known as the Amerithrax terror event, after the FBI's name for the case. Small quantities of pure, powdered spores of anthrax—a natural bacterium that has been developed into a very powerful bioweapon—were placed in envelopes and mailed to several media organizations and to the offices of two United States senators. Five people died after inhaling the spores, while others became critically ill; some of the survivors have never fully recovered. For the most part the victims, including African-Americans and recent immigrants to the United States, were low-level employees of the post office who were just doing their jobs.

The case remained open and unsolved for almost seven years. Then, on July 29, 2008, a USAMRIID scientist named Bruce E. Ivins committed suicide. He had worked as an anthrax researcher in the USAMRIID labs. In a news conference a week later, officials from the FBI and the United States Department of Justice announced that Bruce Ivins was believed by federal investigators to have been likely the sole perpetrator of the anthrax attacks. Federal prosecutors had been getting ready to charge Ivins with the crimes when he killed himself. As of this writing, the FBI's evidence against Ivins has not been fully released, and was never given to a jury for a verdict. Many questions remain. The investigators still don't know, for example, exactly how the powdered anthrax material was made.

Following the anthrax terror event, security at USAMRIID became astronomically tight. After that, it would have been useless for a journalist to ask to go into the space-suit labs. Back at the time when I was researching *The Hot Zone,* though, there was a slight amount of flexibility in the policy. On certain occasions, the Army *had* allowed untrained or inexperienced visitors to go into hot zones at USAMRIID. Unfortunately, as the commander explained to me, some of these visits had ended badly. People who were not familiar with space-suit work with hot agents had a tendency to panic in Level 4, he said.

In one such an incident, a medical doctor—a visitor—who had apparently never worn a biohazard space suit attended a human autopsy in a Level 4 morgue at the Institute. This hot morgue is called the Submarine. The Submarine is a sealed hot zone with an autopsy room and an autopsy table. The cadaver was believed to be infected with a Level 4 Unknown X virus. During the examination, while the space-suited autopsy team was removing organs from the cadaver, some members of the team noticed that the visiting doctor's face seemed red. As the team members looked at him through his faceplate, they saw that his face was also dripping with sweat. Meanwhile, the outer surfaces of his space-suit gloves and sleeves were smeared with blood from the cadaver.

Reportedly, the man began saying, "Get me out!" Suddenly he tore off his helmet and ripped open his space suit, gasping for breath, taking in lungfuls of air from the hot morgue.

The members of the autopsy team took hold of him and hurried him to an air lock door leading to the exit. They opened the door, pushed him into the air lock. At least one of the team members accompanied him into the air lock. The air lock was closed, and the chemical shower was started.

The way I heard the story, the man stood or sat in the air lock while the chemicals ran down inside his opened space suit. The shower stopped automatically after seven minutes. The chemicals had flooded his suit. Then the team members helped him into the staging area—the so-called Level 3 area—and helped pull him out of his space suit. By this time, he was subdued and embarrassed.

At USAMRIID, people who have had a verified exposure to a hot agent are put into a Level 4 quarantine hospital suite called the Slam-

The United States Army Medical Research Institute of Infectious Diseases (USAMRIID)

Getty Images

mer. The Slammer is a biocontainment unit where doctors and nurses wearing space suits can treat a patient without being exposed to a virus the patient may have. The man who had panicked was a possible candidate for quarantine in the Slammer. Even so, after an immediate review of the incident by a safety team, the Army felt that he did not need to be put in quarantine; there was no evidence that the cadaver had actually been infected with a virus. And the man never got sick.

"We can't predict how someone who's untrained might react in BL-4, so we can't allow you to go in," the commander explained to me.

I still wanted to go into Level 4. But I couldn't see how to get there.

IN NARRATIVE NONFICTION WRITING, taking notes is an essential part of the creative process. We tend to think of a reporter's notes as being a transcript of the words of someone speaking to the reporter. If you who are reading this happen to be a student of journalism, remember that you can take notes about anything. It can be quite useful to jot down observations on any and all details of a person and a scene, including sights, smells, and sounds, as well as the emotional aura of the scene. These kinds of observed details might be called deep notes. Deep notes are a record of the visceral reality in which the characters exist—notes on the soup. Deep notes can be details of how people move their bodies, what they wear, what sorts of tics and gestures they display. I always try to note the color of a person's eyes, and, when possible, I try to observe their hands.

One of the main figures in *The Hot Zone* is Lieutenant Colonel Nancy Jaax, an Army space-suit scientist who specializes in Ebola virus. I met Nancy Jaax during my first visit to Fort Detrick, on a warm spring day. She turned out to be a pleasant, energetic, articulate officer who seemed incredibly committed to her work. I learned that she was a mother of two children, who were then in high school. Her eyes were blue-green and active, with flecks of gold encircling the irises. About fifteen minutes into my first interview with her, I asked her if she had ever had a scary experience with Ebola virus. "Oh, sure," she answered. "That's where you realize that habits can save your life."

"What sort of habits?" I asked.

She explained that when you're working with a hot virus like Ebola, it's essential to constantly check your space suit. A suit can get a hole in it. The person inside the suit might not notice the creation of the breach. Nancy Jaax had been trained to frequently check her space suit for leaks. One day, she was cutting open a dead Ebola-infected monkey, and her space suit was splashed with Ebola-infected monkey blood up to the forearms. Then, during a routine safety check, she discovered a hole in the arm of the suit, near the glove. Ebola-infected blood had run down into the hole and was oozing around inside her space suit and had soaked her arm and wrist. "I had an open cut on my hand, with a Band-Aid on it," she said. She'd gotten the cut opening a can of beans for her children. The incident "fell into the category of a close call," she said. In the end, she survived her encounter with Ebola only because her habit of checking her space suit for leaks enabled her to get out of the hot zone fast and remove her bloody space suit. Her narrative left me mesmerized.

There's a useful technique for capturing important moments during an interview that I call the delayed note. When someone is saying something powerful, you don't always want to draw attention to the fact that you're writing down their words, because they may pull back and stop talking. So, on rare occasions, I may stop writing. I put down my notebook. I try to get a neutral expression on my face, as if I'm not that interested. Meanwhile, I'm trying to memorize exactly what the person is saying. When I sense that my short-term memory is getting full, I change the subject and ask a question that I expect will result in a dull answer. The person begins giving the dull answer, and I begin jotting delayed notes in my notebook. I'm writing down what the person said moments earlier, while I was not taking notes. (I learned this technique from John McPhee, who teaches an undergraduate writing course at Princeton University called The Literature of Fact. I had taken his course as a graduate student.) So, as Nancy Jaax began to talk about the blood in her space suit, I put down my notes and listened.

This was just the beginning of the research for a key scene in *The Hot Zone*, narrating how Nancy Jaax got a hole in her space suit and Ebola blood flowed inside it. At one point, much later, I spent twenty minutes sitting with Jaax at her kitchen table, taking notes on her hands. I examined her hands minutely, left and right, back and front, staring at them like a palm reader. Hands are a window into character. Jaax kindly submitted to my study of her hands, though I think it weirded her out just a little.

"Where did you get that scar on your knuckle?" I asked.

"Which one? That one? That's where a goat bit me when I was nine," she answered, touching the scar. It had been a goat on her family's farm in Kansas, she explained, and she could still recall how much the bite had hurt.

Deep notes can also be notes on what a person is thinking. Of course, since you can't read minds, you have to ask people what they are thinking or were thinking. After I've written a passage describing a person's stream of thoughts—a type of narrative that fiction writers refer to as interior monologue—I always fact-check it with the person later. I read the passage aloud, usually on the telephone. I ask the person, "Do these sentences accurately reflect your recollection of what was going through your mind at that moment?"

Often, the person answers, "Not exactly," and proceeds to correct what I've written to make it more faithful to their own memory. If it was an especially dramatic, emotional, or terrifying moment, the person can often give a consistent account of what they were thinking and feeling. (Witnesses to crimes often don't accurately remember the facts of what they saw—but they do remember their feelings with clarity.)

After Jaax had realized that Ebola blood was slopping around inside her space suit, she had to make an emergency exit from Level 4. She went into the air lock and stood in the chemical shower, feeling the blood squishing around on her arm and hand.

"Were you thinking you would die?" I asked her.

"No," she replied. Instead, she had been thinking about the fact that she had forgotten to go to the bank to get money for the babysitter who was taking care of her kids that day. If she was infected with Ebola, the authorities would lock her in the Slammer, and who was going to pay the babysitter?

I don't think a novelist would be likely to invent this. And if it appeared in a novel, it might not ring true. Yet Jaax's account of what she was thinking is completely believable because it occurs in a nonfiction narrative. It seems to reverberate with general human truth. It is a statement about mothers, children, and death, and it cut me to the heart when I heard it. I could not have made it up.

In the bloody space-suit scene, when Nancy Jaax emerged from the chemical shower and took off her suit to examine her hand, to see if there was any Ebola blood on it, I described her hands in detail as her gloves came off. Just a couple of sentences. These sentences were the result of the long examination of her hands at her kitchen table. I mentioned the scar on her knuckle and that she'd gotten it as a girl from a goat bite at her family's farm in Kansas. The scar was a microstory. It told the reader that Nancy Jaax was a Kansas farm girl; she was Dorothy in Level 4.

While I often take photographs to supplement my notes, I almost never use a tape recorder. Apart from the fact that a tape recorder always seems to fail when it's most needed, the device makes the person who's being recorded self-conscious. An interviewee will stare at the tape recorder while his or her speech becomes awkward, not like the natural, lively voice of a person in real life. Indeed, cameras and sound recorders aren't sufficient for deep notes. No electronic recording device can capture the interplay of the human senses. For capturing sensual reality, it seems that only an old-fashioned reporter's notebook has sufficiently advanced technology. I take notes in longhand in little spiral notebooks. They are small enough to fit in a shirt pocket. I use a mechanical pencil.

TWO ARMY VIRUS RESEARCHERS walked down one of the long corridors of USAMRIID, at Fort Detrick. I had been hanging around with them that day. My shirt pocket had a magnetic security card clipped to it, with the word VISITOR on it. My shirt pocket also contained a small reporter's notebook and a mechanical pencil. The walls of the corridor were cinder block, painted the color of sputum. Thick glass windows looked into sealed laboratory suites—Biosafety Level 2, Level 3, and

Level 4. The corridor was filled with a weird, burning, moist smell. This smell, which will never leave my mind, was the odor of giant autoclaves—huge, pressurized steam ovens, full of equipment and waste from the labs, cooking the stuff at high heat and pressure, to make it sterile.

The researchers' names will be given here as Martha and Jeremy; those are not their real names.

Jeremy glanced at Martha. "I was thinking it might be a good day for going into BL-4," he said.

"I was thinking so, too," she answered. She was looking at me.

Jeremy turned to me and said, casually, "Are you up for BL-4 today?"

We were standing before an inconspicuous door marked AA-5. It led to one of the Ebola suites, a group of rooms where Ebola research was actively being done. "What will we be doing in there?" I asked.

"Somebody died," Jeremy answered ominously. "The blood samples came in the other day. We're doing tests to try to identify any virus in the samples."

The dead person's identity had not been disclosed to the Army researchers. They had been told he was a "John Doe" male. He had apparently been a U.S. government employee. He might have been a soldier. Or he might have been a diplomat. Or, possibly, John Doe had worked for an intelligence agency such as the CIA. The researchers would never learn his name, the circumstances of his death, or where he had died. John Doe had died with symptoms that suggested a hot virus. His death had apparently involved tiny, starlike pools of blood under his skin, and blood might have been flowing from his orifices.

"The X virus, if it's there, seems to be Marburg-like," Jeremy said.

Marburg virus is a type of Ebola that is associated with Kitum Cave, in the rain forest on the eastern side of Mount Elgon, an extinct volcano in East Africa. On at least two occasions, people who went inside Kitum Cave died shortly afterward of Marburg virus, having apparently caught the virus inside or near the cave.

"If there's a virus in the blood samples, it's considered a Biosafety Level 4 agent," Jeremy continued. "This is because the guy died and because it [the putative X virus] hasn't been identified. It's an Unknown," he said.

Martha swiped her ID card across a sensor, and a computer-synthesized voice issued from a speaker by the door: "RTU downloading." There was a chime. "One moment please," the computer said. A green light went on, and there was a click. The entry door had unlocked.

Martha went through the entry door and closed it behind her. The light over the door went from green to red.

Meanwhile, Jeremy and I waited in the corridor. Beyond the door was a locker room. Martha would take off all her clothing in the locker room and then would proceed inward through some other rooms, where she would put on her space suit and head into the Ebola zone.

In a few minutes, the light turned green again. This meant that Martha had left the locker room and had gone farther inward. Now we could enter.

Jeremy swiped his security card on a pad by the door, I swiped mine, and we entered the locker room. As we went into the room, we passed through an additional security system that won't be described here.

In the end, I never found out what role the commander had in my visit to Level 4. But I think he must have known about it. Jeremy and Martha were experienced, cautious professionals with impeccable safety records. I speculated that the commander had decided to permit me to visit Level 4 with these two people, so long as the visit was kept low-key—though not unofficial. The main building of USAMRIID had security guards who sat in a control room, monitoring the movements of people in the building. I guessed that the security people had known that I would be going into a hot zone even before I knew it myself. The magnetic security card that I wore clipped to my shirt could show the security staff my location in the building; it would also tell them that I was entering the hot area.

The locker room was small and gritty-looking, with a wooden bench and some steel gym lockers, banged up and dented. Jeremy and I stripped down to nothing and put our clothing in the lockers.

"You need to take off your glasses and leave them, too," he said. "They'd fog up inside the suit, anyway."

People had to take out contact lenses, as well. They also had to remove jewelry, including engagement rings. (Women were allowed to wear a tampon in Level 4, but they had to remove it on their way out of the lab.)

I put my glasses in the locker. As I took off my shirt, I slipped the notebook out of my pocket, and I clipped the mechanical pencil to the notebook. Holding these two items, but otherwise stark naked, I followed Jeremy through a doorway that led inward.

The doorway was actually a tiled shower stall, with a water showerhead in it and a drain in the floor. The shower stall glowed deep purple with ultraviolet lights. On their way out of the hot zone, after having worked in a space suit with dangerous organisms, the researchers would pass naked through the water shower, washing their skin with detergent and bathing their bodies in ultraviolet light, before they went into the locker room to get dressed.

On the inward side of the ultraviolet doorway was a small room considered to be at Biosafety Level 2—the mildest degree of biocontainment. Some metal shelves stood against one wall, and there was a toilet. The shelves were piled with blue cotton surgical scrub suits and other items of surgical wear, along with rolls of tape. We put on scrubs—no underwear was allowed. We also put on white socks, surgical hair coverings, and surgical gloves. Jeremy showed me how to tape the socks and the gloves to the cuffs of my scrub suit, wrapping the tape around the cuff to make a seal. "It's a good idea to empty your bladder," Jeremy said. "You won't get a chance to once you're in a suit."

I took up my notebook and pencil, and we pushed through a door into Level 3—this was a staging area leading to Level 4. It was a large room with cinder-block walls and a concrete floor. We heard a rumble of air-filtering machinery overhead. The staging area was crowded with pieces of laboratory equipment. Along the left-hand wall hung an array of biohazard space suits. They were made of soft blue plastic, with flexible plastic helmet hoods. Most of the suits had the owners' names written on them with a Magic Marker.

"People are touchy about their suits," Jeremy remarked. "It's like the Three Bears: 'Somebody's been wearing my suit.' "

Martha had already donned her space suit, and she was preparing things in the staging area. While Jeremy got into his suit, Martha took a space suit down from the wall and helped me put it on. She opened the suit, peeling apart a sort of large Ziploc-style zipper that ran across the chest, and handed the opened suit to me.

There was no name on my space suit.

I sat down on a bench and slipped my feet through the chest opening, down into the suit's legs. My feet ended up inside a pair of soft plastic bunny feet attached to the suit's legs. I drew the suit over my body, and slid my arms into the sleeves, and into heavy rubber gloves attached to the ends of the sleeves.

I pulled the hood and faceplate down over my head. The space suit's zipper ran diagonally across the chest. I sealed the zipper, running my thumb and fingers along the seal, which closed me into the suit. The zipper was coated with a kind of grease that felt like Vaseline. The seals made a lip-smacking sound as I squeezed the zipper shut.

A number of bright yellow air hoses dangled from the ceiling. They were the source of clean air for people's space suits. Each room in the hot zone had its own set of air hoses hanging from the ceiling. When you moved from one room to another, you had to detach your air hose and get connected to a different air hose in the new room.

I reached up, grabbed an air hose, pulled it downward (it came down from a spring coil). I plugged the hose into a socket near my waist.

My suit inflated with a roaring sound. Cool, dry air began flowing around my body. The air was constantly being bled from exhaust ports in the suit, while the suit remained inflated. I felt the air blowing downward past my face, coming from vents inside the helmet. The airflow was designed to keep the suit under constant positive pressure, so that if the fabric developed a hole or a breach, air would flow *out* of the suit, and wouldn't allow any dangerous organisms to flow in.

The high flow of air inside the suit created an almost deafening roar, which made it difficult to hear anything. The researchers had to shout to carry on a conversation in Level 4.

Jeremy closed his suit and pressurized it.

In front of us was a stainless steel door. It had a biohazard symbol on it, dark red, and the words

BL-4 AREA. BIOHAZARD.
NO UNAUTHORIZED ENTRY.
FULL SUIT PROTECTION REQUIRED.

Martha took a hose and socketed it into her suit, pressurizing it. Then she turned and faced me. She placed herself between me and the steel door. Our suits bumped together. She pushed herself up against me until our faceplates made contact. "HOW ARE YOU DOING?" she asked. It was hard to hear her voice over the roar of the air.

"I'M FINE."

She looked into my eyes. I thought she was examining me for any signs of claustrophobia—flushed skin, sweat standing on my face, wide staring eyes.

Evidently I looked all right. She caught Jeremy's eye and stepped aside.

I took up my notebook and pencil, holding them awkwardly in one of my suit gloves, and we faced the steel door.

Martha looked down at my glove. "WHERE ARE YOU GOING WITH THAT NOTEBOOK?"

"IN THERE," I shouted, indicating the door leading to Level 4.

She began laughing behind her faceplate. "IT'S FINE TO BRING YOUR NOTEBOOK INTO BL-4, BUT IT WILL NEVER COME OUT AGAIN. IT WOULD HAVE TO BE AUTOCLAVED"—cooked in a pressure oven. "THE PAPER WOULD DISSOLVE."

So I couldn't take notes. We would be in Level 4 for at least an hour. I wouldn't be able to recall nearly enough. "I NEED SOME WAY OF TAKING NOTES. WHAT AM I GOING TO DO?"

"YOU COULD TRY THIS." She handed me a sheet of flexible white plastic material, rather like paper, but, she explained, it was coated with Teflon. The researchers used this material in place of paper in a hot zone. You could write on it with a pen. It could be sterilized without damage.

"WHAT ABOUT MY PENCIL?" I asked. "CAN I TAKE THAT IN?"

"UH-UH," Jeremy said, inspecting my mechanical pencil and shaking his head. "THAT'S GOT A SHARP TIP. PUNCTURE YOUR SUIT. SHE'LL GIVE YOU SOMETHING TO WRITE WITH WHEN WE GET TO THE HOT SIDE."

When you moved from one room to another, you had to detach your air hose momentarily, walk without pressure in the suit, and then plug into the next air hose. As soon as the air hose was disconnected, your space suit would deflate.

Martha unclipped her air hose, and her suit went limp around her, shrinking down. She opened the steel door and stepped into the air lock and closed the door behind her. A moment later, Jeremy unclipped his air hose, and his suit went limp, and he stepped through the air lock door.

This left me alone momentarily in the staging room inside a pressurized suit. I looked back at the door we had come in through. I could do it. I could just leave. Open my space suit and climb out of it, and go back to the locker room and put my clothes on. These people wouldn't care; they'd probably be relieved to see me go.

I unplugged my air hose. My space suit lost pressure and collapsed around me. Things got quiet: my air wasn't running. I looked at the exit. Then I turned and opened the steel door and walked into the air lock. The door snapped shut behind me. At the far side of the air lock there was another steel door. I opened it and entered Level 4.

I FOUND MYSELF STANDING in a narrow corridor with walls made of cinder blocks. Martha and Jeremy weren't in sight. I needed air. I looked up and saw a rack of air hoses hanging from the ceiling. I reached up and plugged a hose into my suit. The roaring noise began again, and a rush of air filled my suit, the suit tightening with the pressure. I looked around.

Several pairs of Christopher Robin–style rubber boots were lined up next to the air lock door. You were supposed to wear these boots in Level 4, to protect the bunny feet of the space suit from tearing as you walked around. I stepped into a pair of the boots, feeling nervous. The walls and floor were covered with thick-looking, beige paint—the paint was actually a layer of plastic resin that coated the inner surfaces of the hot zone, like the lining of a swimming pool, keeping it waterproof and airtight.

Martha appeared, coming around a corner. She plugged herself into an air hose. Then she handed me a Bic ballpoint pen. It was a Level 4 hot pen, and it would never leave Level 4, except as a melted lump.

Then we unplugged our air hoses and walked down the corridor. The rooms in Hot Suite AA-5 were small and cramped. They held typical laboratory equipment, the sort that you might find in any medical research lab in a university or corporation. I was struck by the ordinariness of Level 4. There were refrigerators, centrifuges, cabinets stocked with lab supplies. A couple of computers sat on counters. Nothing about the place suggested the presence of dangerous viruses, except for the unforgettable fact that we were inside space suits. There were sinks with water faucets. Wastewater from the sinks ran into collection tanks where the waste was sterilized. There was a room that had nothing in it except for two large chest freezers. They were hot freezers, hot as hell; I wondered if they contained Ebola, but I didn't dare ask. The freezers were locked and were equipped with alarms. They could be opened only with combination keypads.

I followed Jeremy and Martha around a corner into a small room, where we plugged in our air hoses. Shelves and counters ran around the walls. A sign said:

BIOHAZARD

HAZARD IDENTITY: ____________

The line had been left blank, since that day the researchers were working with an Unknown.

On one counter sat an incubator—a metal box that was kept warm inside, at a temperature of 37 degrees Celsius, which is 98.6 degrees Fahrenheit. The temperature of a human body. Martha opened the incubator and removed a rack full of small plastic flasks containing the

Unknown. The flasks were sealed tightly with screw caps and contained a pinkish fluid. The fluid was a nutrient bath for a layer of living cells from the kidney of a monkey that were growing on a flat inner surface of each flask. Tiny drops of the blood of victim John Doe had been introduced into these flasks. If Doe's blood contained a virus, the virus would be likely to infect the monkey cells. Then the cells would start dying—shriveling up and bursting—and this would be evidence that the flask contained a virus that had come originally from John Doe's blood.

As yet, there was no direct evidence that John Doe had been infected with a virus. He might have died from some other cause. (Other illnesses, such as malaria, can mimic some of the symptoms of Ebola or Marburg, but we had to assume that John Doe had already been tested by doctors who'd ruled out the more common infections and illnesses.) It was the job of the researchers to determine whether the blood contained a virus and, if so, to try to identify it. This was detective work. The procedure was that of Sherlock Holmes: you rule out possibilities until only one possibility remains.

Martha wanted to look at the monkey cells to see if any of them showed signs of being infected with a virus. She sat down at a counter and placed one of the little flasks under a microscope. She did not open the flask. She stared into the eyepieces of the microscope through her faceplate, turning her head back and forth. "DO YOU KNOW HOW HARD IT IS TO GET YOUR HEAD COCKED SO YOU CAN SEE INTO A MICROSCOPE?" she yelled over the unending roar of air. She stood up. "DO YOU WANT TO LOOK?"

I sat down on the chair before the microscope. I had difficulty seeing through my faceplate into the eyepieces. I began twisting my head around. Finally I got a clear look through the microscope into the flask.

I could see glittering fields of monkey cells, reddish gold in color. Were they sick or healthy? They were just cells to me. As for the X virus, if any particles of it had been present in the cells they would not have been visible through the microscope. Virus particles are too small to be seen with an optical microscope. Most viruses can be seen only with an electron microscope, which magnifies things that are extremely small. A cold virus particle sitting at the base of an eyelash hair would be like a peanut sitting by the Washington Monument.

As I was staring into the eyepieces of the microscope, I felt a popping sensation. Something felt weird around my chest. My suit began to feel sloppy and loose. But I was interested in the cells and wasn't paying attention to my space suit. Eventually, though, I moved away from the microscope and stood up. That was when I realized that my space suit had blown open in the hot zone.

AIR WAS RUSHING OUT around my neck and chest, pouring out of the suit. With a growing sense of alarm, I looked down to try to see what was happening, but couldn't see anything. The lower part of my helmet hood blocked the view. I started feeling around with my gloves, but couldn't get much sensation, for the gloves were thick and clumsy. I began tugging on something, some sort of fabric. To my horror, I realized that I was grabbing at my surgical scrubs. I was feeling around *inside* my space suit.

I knew what had happened. As I had been bending over the microscope, the movement had twisted the zipper that ran across the suit's chest, and the lips of the zipper had pulled apart, and the pressure in the suit had opened it completely. And now I couldn't get the zipper closed. I almost threw a tubular cast.

"DO I HAVE A PROBLEM HERE?" I yelled.

Jeremy had been working with his back toward me. He swung around, looked at me, and swore. He moved toward me, holding his palms outward. He ran his palms back and forth on my chest, closing the zipper.

My suit swelled up and tightened, regaining pressure.

"HOW BAD WAS THAT?" I asked.

He looked a little embarrassed. "THE ZIPPERS GET WORN. THEY CAN POP OPEN."

I had ended up with a ratty old piece of Army gear, a space suit that belonged to nobody. A little voice started speaking in my head. *What are you doing here?* the voice said. *You're in an Ebola lab in a fucking defective space suit.* I started to feel giddy. It was an intoxicating rush of fear, a sensation that all I needed to do was relax and let the fear take hold, and I could drift away on waves of panic, screaming for help.

Martha was looking into my eyes again.

The little voice went on: *You're headed for the Slammer.*

Jeremy tried to soothe me. He assured me that the incident was not, in fact, an exposure to a hot agent. The suits did occasionally pop open, he admitted. "THE THING IS, YOUR SUIT HAD POSITIVE PRESSURE THE WHOLE TIME," he explained. "THE AIR WAS FLOWING DOWN PAST YOUR FACE AND OUT OF THE OPENING. NOTHING COULD GET INTO YOUR SUIT. IT COULDN'T MOVE UP TO YOUR FACE."

I really wanted to believe that an Unknown virus was not having an encounter with me inside my space suit. The air had been gushing out, I told myself. The flow would have carried any particles of a hot agent out of my suit. Anyway, if something had gotten in, by now it was too late to panic. I told the scientists that I wanted to remain in Level 4. After that, I touched my chest zipper frequently to make sure it was closed. *Habits can save your life.*

The researchers had work to do. Martha intended to open a flask of the Joe Doe Unknown. She carried one of the flasks to a Steriguard safety hood—a cabinet that produced a curtain of air blowing between the virus samples and the person sitting there. This air curtain acted as a shield, preventing any drifting particles from coming near the researcher. The cabinet also had a sliding glass door. If you opened the glass too far—potentially releasing a virus into the air of the hot zone—an alarm would go off.

With a pair of tweezers, Martha picked up a glass slide that had spots of reactive compound on it. It was called a spot slide. You drop liquid samples of virus on the spots, and if a spot changes color, it helps you identify the virus. "WE HANDLE GLASS WITH TWEEZERS BECAUSE YOU DON'T WANT TO PICK UP ANYTHING MADE OF GLASS WITH YOUR FINGERS," she said. "YOU NEVER WANT TO CUT YOURSELF IN HERE."

Martha opened the flask containing the Unknown. She took up a push-button pipette—a device that is used in biological labs for moving very small quantities of fluid from one place to another.

She inserted the tip of the pipette into the open flask and pushed a button, and the pipette sucked up a small amount of the liquid containing the Unknown X. She positioned the pipette over the glass slide and placed droplets of the liquid on the slide. Her hands moved with deft precision.

"DO YOU WANT TO TRY DOING THIS?" she asked.

I sat down in front of the safety cabinet. She handed me another flask. I had difficulty removing the cap from the flask. My heavy rubber space-suit gloves were impossible. I picked up the pipette and began dropping the liquid into a row of small test tubes, so that tests could be done on it. Even this simple task proved to be achingly slow and difficult. I couldn't see how anyone could do medical research wearing Mickey Mouse gloves and a space suit. The fact that the Army researchers were able to do it every day made me appreciate the depth of their skill and training. My cheek began to itch, but I couldn't figure out how to scratch it, since my head was inside a helmet hood. When I had finished the task, I closed the flask and held it up to the light.

The liquid shimmered inside; I was face-to-face with a presumed Level 4 Unknown. My cheek was itching badly. "HOW DO YOU DEAL WITH AN ITCH IN ONE OF THESE SUITS?" I asked.

"YOU DON'T," Jeremy shouted above the unending roar of air.

FINALLY IT WAS TIME to make an exit from the hot zone. I followed the researchers through the maze of corridors and little rooms to the air lock. Jeremy entered the air lock first and started the chemical shower by pulling on a chain. The shower began running in the air lock, sterilizing the outside of his space suit. While Martha and I waited for Jeremy to finish his chemical shower, I handed her my Level 4 Bic pen. She left it near a computer for the next person to use. Then Martha and I went into the air lock together. She pulled the chain and we stood under the chemical shower. The chemicals gave off a strong but not unpleasant smell, which eventually crept inside my space suit.

Martha pointed to my Teflon paper—my notes—which I held clutched in my glove. "LET ME HAVE THAT FOR A SEC," she said. She crumpled it up, dipped it into a bath of chemicals, and then, using both hands, she scrubbed the Teflon paper against itself and squeezed it, as if she were rinsing a washrag. After a minute or so, she pulled the paper out of the chemical bath. My notes were wrinkled, wet, and sterile. The shower stopped, and I opened the steel door and stepped into the normal world, holding the notes.

Later, I wrote about Nancy Jaax's feelings after she had gotten a hole in her space suit and she was standing in the chemical shower, feeling Ebola blood oozing around inside her suit and wondering who was going to pay the babysitter. I constructed the passage primarily from detailed interviews with Nancy Jaax, of course. Yet there is something else in that scene that did not appear in the book. It was an iceberg of personal experience, one I hadn't felt able to write about until now. I had been in the rooms she had been in. There, I had experienced a breach condition in my space suit, too, and it had happened in the presence of a putative hot Marburg-like Unknown. And I had stood in the same chemical shower afterward, with thoughts and fears pouring through my mind. . . . I had been boiled in the soup.

I LOVE EXPLORING UNSEEN WORLDS. In this book, we are embarking on a deep probe through the realms of the vanishingly small, where, at times, all we can say is "There be monsters." The chapters in this book were originally published in *The New Yorker,* but I've expanded, updated, and linked them.

One monster of the microscopic universe is a mysterious genetic disease, called Lesch-Nyhan syndrome, which is caused by the alteration of a single letter of a person's DNA code. If one letter of the human DNA is altered in a certain place in the code, the person who is born with the tiny error has a dramatic change of behavior—a lifelong, irresistable compulsion to attack himself, chewing off . . . it's in the last chapter.

"The Mountains of Pi" describes David and Gregory Chudnovsky, mathematicians who built a supercomputer out of mail-order parts in Gregory's apartment in New York City. They were using their homemade supercomputer to calculate the number pi (π) to billions of decimals. They were looking deep into pi, down into an infinitesimal smallness of precision, deeper and deeper into pi, trying to get a glimpse of the face of God.

I originally wrote about the Chudnovskys in a "Profile" for *The New Yorker.* When I first met them and began researching the piece, they seemed pleased that I was writing about pi, but they soon got the idea that I was also writing about *them*. They began to object. "My dear fellow, can't you leave our names out of this?" David said.

I had to explain that it is not really feasible to leave a person's name out of a *New Yorker* "Profile" of him.

This puzzled me, why the Chudnovskys didn't want their names used. The answer, as I finally figured out, had to do with the nature of mathematics as a human activity. Mathematics is not strictly science, nor is it absolutely art. Mathematics is both objectively rigorous and highly creative, and so it spans the divide between the two worlds, and expresses the unity of science and art. In effect, mathematics is a cathedral of the intellect, built over thousands of years, displaying some of the greatest achievements of the human spirit. The Chudnovsky brothers saw themselves as anonymous workers adding a few details to the cathedral. Their names didn't matter.

When I had finally gotten their reluctant assent to let me write about them as people and had finished drafting "The Mountains of Pi," a fact-checker from *The New Yorker* named Hal Espen paid a visit to the Chudnovsky brothers in order to verify the facts in my piece. Soon afterward, Gregory phoned me in a state of indignation. Hal Espen had spent a long time in Gregory's apartment, looking at the things I'd described and asking the brothers many questions. At one point, he wanted to confirm that Gregory owned hand-sewn socks made of scraps of cloth, as I had written, so he asked Gregory if he could see his socks. Gregory wasn't wearing them, so Espen ended up looking in one of the drawers of Gregory's dresser, where he found the socks and verified my description of them. "He was a nice guy, but why did he have to rifle through my socks?" Gregory demanded.* If the names of the cathedral workers didn't matter, their socks mattered

*Hal Espen later became a senior editor at *The New Yorker* and later the editor of *Outside* magazine.

even less. But Gregory Chudnovsky's socks mattered deeply to me, for the same reason that the color of Nancy Jaax's eyes and the scar on her hand mattered. The business of a writer, in the end, is human character, human story. Unlike a novelist, a narrative nonfiction writer cannot make up details of character. For the nonfiction writer, details must be found where they exist, like diamonds lying in the dust, unnoticed by passing crowds.

In exploring biology, I moved my focus away from the smallest life-forms—viruses—and began climbing the coast redwood trees of California. The coast redwoods are the largest individual living things in nature. Redwoods can be nearly forty stories tall; they would stand out in midtown Manhattan. In order to climb a redwood, you put on a harness and ascend hundreds of feet up a rope into the redwood canopy. It's like scuba diving, except that you go into the air. The canopy is the aerial part of a forest, and it is an unseen world, invisible from the ground. Once you have ascended into the redwood canopy, which is the world's tallest canopy, you dangle in midair, in a harness, around thirty stories above the ground. You are suspended from ropes attached to branches overhead. You move through the air while hanging on ropes, sometimes going from tree to tree. The redwood canopy is a lost world, unexplored, out of sight, teeming with unknown life. After writing a book about it (*The Wild Trees*), I learned of the existence of an unexplored rain-forest canopy in eastern North America; I hadn't known there were rain forests in the East. This eastern rain forest was being destroyed by parasites invading the ecosystem. It was an unseen world that was vanishing even before it had been explored by humans. Thus an interest in giant life-forms ended up returning me to a focus on small things: "A Death in the Forest."

"The Blood Kiss" is about the search for the unknown host of the Ebola virus, and it narrates an outbreak of Ebola in Congo. In researching it, I ended up talking with a medical doctor named William T. Close. Bill Close, who is the father of the actress Glenn Close, had been the head of the main hospital in Kinshasa, the capital of Congo (then called Zaire). Then, while I was staying late in the offices of *The New Yorker* on a Friday night—closing time, when the magazine is put to bed—I telephoned Dr. Close for a last-minute fact-checking conversation. In about ten minutes (I had been told) the magazine would be closed—finished—and would be transmitted electronically to the printing plant in Danville, Kentucky. That was when Dr. Close told me the story of a Belgian doctor who had performed a terrifying act that could be called a blood kiss, with a patient in an Ebola ward.

"Oh, my God," I blurted.

I asked someone to go find Tina Brown, the editor, and see if the magazine could be held open for a little while, as a doctor was saying something. It was okay with Brown, but there was no time to take notes. I asked Close to tell me the story again, while I scrawled sentences describing the blood kiss on a sheet of paper. With Close waiting on the line, I carried the paper over to the make-up department (an office where the magazine's compositors worked at computer screens) and I handed it to Pat Keogh, the head of make-up, who typed my scrawl into the master electronic proof. Almost immediately, the passage was reviewed for grammar by *The New Yorker*'s grammarian, a quiet person named Ann Goldstein. (Experts who are being quoted in *The New Yorker* are encouraged not to use bad grammar.)

Minutes later, with Bill Close still waiting on the telephone, we heard that his grammar was okay, and we heard that the magazine had been transmitted electronically to the printing plant. I said good night to the now-eponymous Dr. Close, and hung up.

That was when something struck me. How could I have been so stupid? I had forgotten to ask Close *what had happened to the Belgian doctor afterward*. He would have died a grisly death. Years later, I learned what had transpired, and you will read it here.

In "The Human Kabbalah," I explored the decoding of the human DNA and the resulting fantastic stock-market bubble that made the genomic scientist J. Craig Venter a billionaire for a while. One day, while working on this story, I was hanging out in the laboratory of the Nobel laureate Hamilton O. Smith—one of the great figures in the history of molecular biology—and I mentioned to him that I was having trouble, as a writer, describing DNA in a physical sense. I wanted readers to get a concrete picture of it in their minds. "The trouble is, DNA is invisible," I said to Hamilton Smith.

"No, it isn't," he said. He asked me if I'd ever seen it. I hadn't, so he dribbled some purified DNA out of a test tube. It looked like clear snot.

"What does it taste like?" I asked him.

That surprised him a little. He didn't know. In almost forty years of research with DNA, Hamilton Smith had never tasted the molecule.

As soon as I got home, I ordered some pure, dried DNA through the mail. It arrived: a bit of fluff in a bottle. I put some of it on my tongue. Sure enough, it had a taste. In taking notes, it is useful to remember that all the senses can be involved.

I continued to follow the Chudnovsky brothers and their journeys in the universe of numbers. This led them, with me trailing behind, to the mysterious Unicorn Tapestries, owned by the Metropolitan Museum of New York. All the while, I was researching the curious genetic disease that causes people to mutilate and even, in effect, cannibalize their own bodies—Lesch-Nyhan syndrome.

The disease was almost unbearable to contemplate, and at first almost impossible to describe. It made its victims seem inhuman. I couldn't find a way into the writing of the story, despite spending months and finally years on it. Ultimately, the last chapter of this book took me seven years to write. The disease probably could not be invented by a fiction writer, or if it were invented, it would not seem believable. Yet there it was, an undeniable reality. I needed to understand, if possible, what it might feel like to have this disease. I wanted to try to connect the seemingly unknowable experience of self-cannibalism to that of common humanity. Henry Fielding, in his famous preface to his novel *The History of Tom Jones,* in which he defined the basic terms of fiction writing, quoted Terence: "I am human: nothing human is alien to me." If people with Lesch-Nyhan disease were human, then they could not be alien to us. The only way to find humanity in the story was to climb into the soup with two people who had been born with the disease and start taking notes.

Panic in Level 4

The Mountains of Pi

WHEN HE WAS THIRTY-SIX, Gregory Volfovich Chudnovsky began to build a supercomputer in his apartment from mail-order parts. Gregory Chudnovsky was, and is, a number theorist, a mathematician who studies numbers, and he felt that he needed a supercomputer to do it. His apartment was situated near the top floor of a run-down building at the corner of 120th Street and Amsterdam Avenue, on the West Side of Manhattan. Around the time he decided to build the supercomputer, a corpse was found stuffed into a garbage can at the end of his block. The project officially took two years, though in reality it never ended. At the time he began the project, the world's most powerful supercomputers included the Cray series, the Thinking Machines arrays, the Hitachi line of supercomputers, the nCube, the Fujitsu machine, the Kendall Square Research machine, the NEC supercomputer, the Touchstone Delta, and Gregory Chudnovsky's apartment. The apartment was a kind of container for the supercomputer at least as much as it was a container for people.

Gregory Chudnovsky's partner in the design and construction of the supercomputer was his older brother, David Volfovich Chudnovsky. ("Volfovich" means "Son of Wolf.") David was, and is, also a number theorist, and he lived five blocks away from Gregory. The Chudnovsky brothers were reluctant to give a name to their machine. To them, it was a household appliance that could help with their investigation of numbers. You didn't give a name to your toaster oven, so why would you give a name to your supercomputer?

When I pressed the Chudnovsky brothers to give me some sort of a name for it, they shrugged and said it was nothing.

"I don't want to call it nothing," I said to the brothers.

"Why not?" David answered. However, he said, as a convenience I could refer to it as "m zero."

At any rate, the "zero" in the machine's name hinted at a history of failures—three previous duds in Gregory's apartment, three homemade supercomputers that hadn't worked. The brothers referred to these machines as negative three, negative two, and negative one. The brothers broke them up for scrap, and they got on the telephone and ordered more parts.

Whatever the supercomputer was, it filled the former living room of Gregory's apartment, and its tentacles reached into other rooms. The brothers claimed that m zero was a "true, general-purpose supercomputer" and that it would turn out to be as fast and powerful as a Cray Y-MP. A Cray Y-MP had a sticker price of more than thirty million dollars. A Cray was a black cylinder seven feet tall, and it was cooled by liquid freon. The brothers spent around seventy thousand dollars on parts for their supercomputer, and much of the money came out of their wives' pockets. Seventy thousand dollars was a little more than two-tenths of one percent of the cost of a Cray.

It was safe to say that Gregory Chudnovsky was one of the world's leading architects of supercomputers. He had an ability to see the design of a supercomputer in his mind's eye. He liked to imagine supercomputers that might never be built, like an architect who dreams of towers and cities in a splendid future. M zero was incredibly fast. Gregory called it a relativistic machine, because he had woven the design of the machine around Einstein's theory of special relativity. M zero's network of processors shuttled numbers around it so fast that the different parts of the machine operated in slightly different space-times.

Gregory Chudnovsky had a spare frame and a bony, handsome face. He had a long beard, streaked with gray, and dark, unruly hair, a wide forehead, and wide-spaced brown eyes. He walked in a slow, dragging shuffle, leaning on a bentwood cane, while his brother, David, typically held him under one arm, to prevent him from toppling over. He had myasthenia gravis, an autoimmune disorder of the muscles. The symptoms, in his case, were muscular weakness and difficulty in breathing. "I have to lie in bed most of the time," Gregory told me. His condition seemed to be getting gradually worse. He developed the disease when he was twelve years old, in the city of Kiev, Ukraine, where he and David grew up. In those days, Ukraine was part of the old Soviet Union. Now Gregory spent his days sitting or lying in a bed heaped with pillows, in his bedroom down the hall from the room that housed the supercomputer. Gregory's bedroom was filled with paper. It contained, by my estimate and the calculation of a *New Yorker* fact-checker, at least one ton of paper. He called his bedroom his junkyard. The room faced east. It would have been full of sunlight in the morning if he'd ever raised the shades, but he kept them lowered, because light hurt his eyes.

You almost never met one of the Chudnovsky brothers without the other. You usually found the brothers conjoined, like Siamese twins, David holding Gregory by the arm or under the armpits, speaking to him tenderly, cautioning him to be careful not to fall or hurt himself. They worked together so closely that they claimed to be a single mathematician who by chance happened to occupy two human bodies. They completed each other's sentences and interrupted each other, but they didn't look completely alike. While Gregory was thin and bearded, David was portly, with a plump, clean-shaven face. David's manner was refined and aristocratic. Black-and-gray curly hair grew thickly on top of his head, and he had heavy-lidded pale blue eyes, which had a melancholy look. He always wore a starched white shirt and, usually, a muted silk necktie. His tie rested on a bulging stomach.

The Chudnovskian supercomputer, m zero, burned two thousand watts of power. It ran day and night. The brothers didn't dare shut it down; they were afraid it would die if they did. At least twenty-five fans blew air through the machine to keep it cool; otherwise something might melt. Waste heat permeated Gregory's apartment, and the room that contained the supercomputer climbed to more than a hundred degrees Fahrenheit in the summer. The brothers kept the apartment's lights turned off as much as possible. If they switched on too many lights while m zero was running, they feared they might start an electrical fire. Gregory couldn't breathe city air without developing lung trouble, so he kept the apartment's windows closed all the time. He had air conditioners running in them during the summer, but that didn't seem to reduce the heat. As the temperature climbed on hot days, the inside of the apartment smelled of cooking circuit boards, a sign that m zero was not well. A steady stream of boxes arrived by Federal Express, and an opposing stream of boxes flowed back to mail-order houses, containing parts that had overheated, failed, bombed, or acted strange, along with letters from the brothers demanding an exchange or their money back. The building superintendent didn't know that the Chudnovsky brothers were using a supercomputer in Gregory's apartment. The brothers were afraid he would find out.

The Chudnovskys, between them, had published more than a hundred and fifty papers and twelve books, mostly on the subject of number theory or mathematical physics. They lived in Kiev until 1977, when they left the Soviet Union and, accompanied by their parents, went to France. The family lived there for six months, where David fell in love with a French diplomat named Nicole Lannegrace, and they were married. The Chudnovsky brothers, along with their parents and Nicole Lannegrace, immigrated to the United States and settled in New York, where Nicole became a diplomat with the United Nations. The brothers eventually became American citizens.

The brothers enjoyed an official relationship with Columbia University: Columbia called them senior research scientists in the Department of Mathematics, but they didn't have tenure, they didn't teach students, and they didn't attend faculty meetings. They were lone inventors, operating out of Gregory's apartment. Gregory's wife, Christine Pardo Chudnovsky, was an attorney with a midtown law firm. She had been an undergraduate at Columbia University when Gregory arrived there, and she'd fallen in love with him at first sight. Nicole Lannegrace's salary as a U.N. diplomat and Christine's as a lawyer helped cover much of the funding needs of the brothers' supercomputing complex in Gregory and Christine's apartment. Gregory and David's mother, Malka Benjaminovna Chudnovsky, a retired engineer, was living with Gregory and Christine and was in poor health. David spent his days in Gregory's apartment, taking care of his brother, their mother, and m zero.

When the Chudnovskys applied to leave the Soviet Union, it attracted the attention of the KGB. The brothers happened to be friends with the physicist Andrei Sakharov, a key inventor of the Soviet hydrogen bomb, who had later become a human-rights activist and a proponent of nuclear disarmament, getting himself into serious trouble with the Kremlin. The Chudnovskys' association with Sakharov, as well as the fact that they were Jewish and mathematical, attracted at least a dozen KGB agents to their case. The brothers' father, Volf Grigorevich Chudnovsky ("Wolf, Son of Gregory") was severely beaten by KGB agents in 1977. Volf died in 1985, in New York City, of what the brothers believed were lingering effects of his torture. Volf Chudnovsky was a professor of civil engineering at the Kiev Architectural Institute, and he specialized in the structural stability of buildings, towers, and bridges. Not long before he died, he constructed in Gregory's apartment a labyrinth of bookshelves, his last work of civil engineering. Volf's bookshelves extended into every corner of the apartment, and they had become packed with literature and computer books and books on history and art and, above all, books and papers on the subject of numbers. Since almost all numbers run to infinity (in digits) and are totally unexplored, an apartment full of writings on numbers holds hardly any knowledge about numbers at all. Numbers, and the patterns of relationships among them, are powerful, deep, and mysterious. It is not at all clear that the human mind evolved in such a way that it is very much able to understand numbers. But it helps to have a supercomputer on the premises to advance the work.

ONE DAY, I called the Columbia University math department trying to find out how to make contact with the Chudnovskys. I had read a short news item about them but could learn very little that was definite. They were reportedly living somewhere in New York City. However, they did not seem to be listed in the Manhattan telephone book, and they didn't have an unlisted telephone number, either. (I learned later that they actually were listed in the Manhattan telephone book but under a nonexistent name.) "The Chudnovskys?" the person who answered the phone at Columbia said. "I have no idea where they are. We haven't seen them around here in a long time. I have an old phone number for them. Somebody said it doesn't work anymore."

I dialed the number and got a fax tone. I handwrote a message on a piece of paper and faxed it, asking if this number belonged to the Chudnovskys and, if so, would they be able to meet with me? There was no reply. Weeks passed. I gave up. But then one day my phone rang; it was David Chudnovsky. "Look, you are welcome," he said. He had a genteel-sounding voice with a Russian accent.

On a cold winter day soon afterward, I rang the bell of Gregory's apartment on 120th Street. I was carrying a little notebook and a mechanical pencil in my shirt pocket. David answered the door. He pulled the door open a few inches, and then it stopped. It was jammed against an empty cardboard box and a mass of hanging coats. He nudged the box out of the way with his foot. "Don't worry," he said. "Nothing *unpleasant* will happen to you here. We will not turn *you* into digits." A Mini Maglite flashlight protruded from his shirt pocket.

We were standing in a long, dark hallway. The place was a swamp of heat. My face and armpits began to drip with sweat. The lights were off, and it was hard to see anything. This was the reason for David's flashlight. The hall was lined on both sides with bookshelves supporting huge stacks of paper and books. The shelves took up most of the space, leaving a passage about two feet wide running down the length of the hallway. At the end of the hallway was a French door. Its mullioned glass panes were covered with translucent paper. The panes glowed.

We went along the hallway. We passed a bathroom and a bedroom door, which was closed. The bedroom belonged to Malka Benjaminovna Chudnovsky. We passed a sort of cave containing vast amounts of paper. This was Gregory's bedroom, his junkyard. We passed a small kitchen, our feet rolling on computer cables. David opened the French door, and we entered the living room. This was the chamber of the supercomputer. A bare lightbulb burned in a ceiling fixture. The room contained seven display screens, two of which were

The Chudnovsky Mathematician: Gregory and David Chudnovsky in Gregory's New York City apartment, 1992.
Irena Roman

filled with numbers; the other screens were turned off. The windows were closed and the shades were drawn. Gregory Chudnovsky sat on a chair facing the lit-up screens. He wore a tattered and patched lamb's wool sweater, a starched white shirt, blue sweatpants, and the hand-stitched two-tone socks. From his toes trailed a pair of heelless leather slippers. His cane was hooked over his shoulder, hung there for convenience. "Right now, our goal is to compute pi," he said. "For that we have to build our own computer." He had a resonant voice and a Russian accent.

"We are a full-service company," David said. "Of course, you know what 'full-service' means in New York. It means 'You want it? You do it yourself.' "

A steel frame stood in the center of the room, screwed together with bolts. It held split-open shells of personal computers—cheap PC clones, knocked wide open like cracked walnuts, their meat spilling all over the place. The brothers had crammed superfast logic boards inside the PCs. Red lights on the boards blinked. The floor was a quagmire of cables.

The brothers had also managed to fit into the room masses of empty cardboard boxes, and lots of books (Russian classics, with Cyrillic lettering on their spines), and screwdrivers, and data-storage tapes, and software manuals by the cubic yard, and stalagmites of obscure trade magazines, and a twenty-thousand-dollar engineering computer that they no longer used. "We use it as a place to stack paper," Gregory explained. From an oval photograph on the wall, the face of Volf Chudnovsky, their late father, looked down on the scene. The walls and the French door were covered with sheets of drafting paper showing circuit diagrams. They resembled cities seen from the air. Various disk drives were scattered around the room. The drives were humming, and there was a continuous whir of fans. A strong warmth emanated from the equipment, as if a steam radiator were going in the room. The brothers were heating the apartment with silicon chips.

"Myasthenia gravis is a funny thing," Gregory Chudnovsky said one day from his bed in his bedroom, the junkyard. "In a sense, I'm very lucky, because I'm alive, and I'm alive after so many years. There is no standard prognosis. It sometimes strikes young women and older women. I wonder if it is some kind of sluggish virus."

It was a cold afternoon, and rain pelted the windows; the shades were drawn, as always, and the room was stiflingly warm. He lay against a heap of pillows with his legs folded under him. His bed was surrounded by freestanding bookshelves packed and piled with ramparts of stacked paper. That day, he wore the same tattered wool sweater, a starched white shirt, blue sweatpants, and another pair of handmade socks. I had never seen socks like Gregory's. They were two-tone socks, wrinkled and floppy, hand-sewn from pieces of dark blue and pale blue cloth, and they looked comfortable. They were the work of Malka Benjaminovna, his mother. Lines of computer code flickered on the screen beside his bed.

This was an apartment built for long voyages. The paper in the room was jammed into bookshelves along the wall, too, from floor to ceiling. The brothers had wedged the paper, sheet by sheet, into manila folders, until the folders had grown as fat as melons. The paper was also stacked chest-high to chin-high on five chairs (three of them in a row beside his bed). It was heaped on top of and filled two steamer trunks that sat beneath the window, and the paper had accumulated in a sort of unstable-looking lava flow on a small folding cocktail table. I moved carefully around the room, fearful of triggering a paperslide, and I sat down on the room's one empty chair, facing the foot of Gregory's bed, my knees touching the blanket. The paper surrounded his bed like the walls of a fortress, and his bed sat at the center of the keep. I sensed a profound happiness in Gregory Chudnovsky. His happiness, it occurred to me later, sprang from the delicious melancholy of a life spent largely in bed while he explored a more perfect world that opened through the portals of mathematics into vistas beyond time or decay.

"The system of this paper is archaeological," he said. "By looking at a slice, I know the date. This slice is from five years ago. Over here is some paper from four years ago. What you see in this room is our working papers, as well as the papers we used as references. Some of the references we pull out once in a while to look at, and then we leave them in another pile. Once, we had to make a Xerox copy of the same book three times, and we put it in three different piles, so we could be sure to find it when we needed it. There are books in there by Kipling and Macaulay. Eh, this place is a mess. Actually, when we want to find a book it's easier to go to the library."

Much of the paper consisted of legal pads covered with Gregory's handwriting. His handwriting was dense and careful, a flawless minuscule written with a felt-tipped pen. The writing contained a mixture of theorems, calculations, proofs, and conjectures concerning numbers. He used a felt-tip pen because he didn't have enough strength in his hand to press a pencil on paper. Mathematicians who had visited Gregory Chudnovsky's bedroom had come away dizzy, wondering what secrets the scriptorium might hold. He cautiously referred to the steamer trunks beneath the window as valises. They were filled to the lids with compressed paper. When Gregory and David flew to Europe to speak at conferences on the subject of numbers, they took both "valises" with them, in case they needed to refer to a proof or a theorem. Their baggage particularly annoyed Belgian officials. "The Belgians were always fining us for being overweight," Gregory said.

The brothers' mail-order supercomputer made their lives more convenient. It performed inhumanly difficult algebra, finding roots of gigantic systems of equations, and it constructed colored images of the interior of Gregory Chudnovsky's body. They used the supercomputer to analyze and predict fluctuations in the stock market. They had been working with a well-known Wall Street investor named John Mulheren, helping him get a profitable edge in computerized trades on the stock market. One day I called John Mulheren to find out what the brothers had been doing for him. "Gregory and David have certainly made us money," Mulheren said, but he wouldn't give any details on what the brothers had done. Mulheren had been paying the Chudnovskys out of his trading profits; they used the money to help fund their research into numbers. To them, numbers were more beautiful, more nearly perfect, possibly more complicated, and arguably more real than anything in the world of physical matter.

THE NUMBER PI, or π, is the most famous ratio in mathematics. It is also one of the most ancient numbers known to humanity. Nobody knows when pi first came to the awareness of the human species. Pi may very well have been known to the builders of Stonehenge, around 2,600 B.C.E. Certainly it was known to the ancient Egyptians. Pi is approximately 3.14—it is the number of times that a circle's diameter will fit around a circle. On the following page is a circle with its diameter.

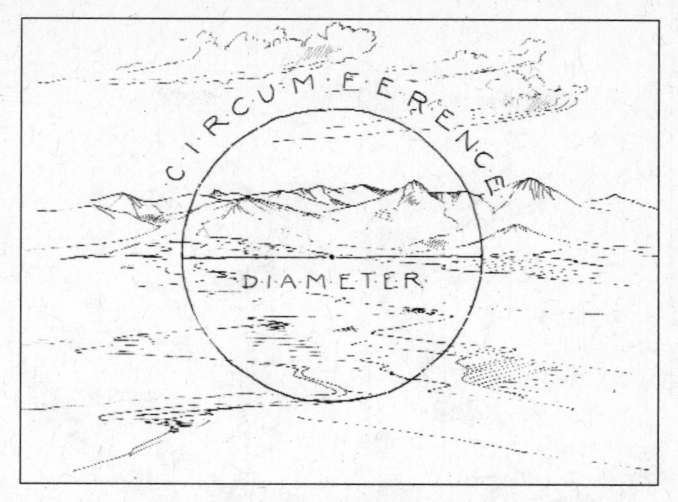

Landscape with a circle and its diameter. This drawing shows a rough visual approximation of pi.
Drawing by Richard Preston

Pi is an exact number; there is only one pi. Even so, pi cannot be expressed *exactly* using any finite string of digits. If you try to calculate pi exactly, you get a chain of random-looking digits that never ends. Pi goes on forever, and can't be calculated to perfect precision: 3.14159265358979323846264338327950288419716939937 51. . . . This is known as the decimal expansion of pi. It is a bloody mess. If you try to express pi in another way, using an algebraic equation rather than digits, *the equation goes on forever.* There is no way to show pi using digits or an equation that doesn't get lost in the sands of infinity. Pi can't be shown completely or exactly in any finite form of mathematical representation. There is only one way to show pi exactly, and that is with a symbol. See the illustration on the following page for a symbol for pi.

The pizza pi I baked and drew, here, is as good a symbol for pi as any other. (It tasted good, too.) The digits of pi march to infinity in a predestined yet unfathomable code. When you calculate pi, its digits appear, one by one, endlessly, while no apparent pattern emerges in the succession of digits. They never repeat periodically. They seem to pop

Pi.
Drawing by Richard Preston

up by blind chance, lacking any perceivable order, rule, reason, or design—"random" integers, ad infinitum. If a deep and beautiful design hides in the digits of pi, no one knows what it is, and no one has ever caught a glimpse of the pattern by staring at the digits. There is certainly a design in pi, no doubt about it. It is also almost certain that the human mind is not equipped to see that design. Among mathematicians, there is a feeling that it may never be possible for an inhabitant of our universe to discover the system in the digits of pi. But for the present, if you want to attempt it, you need a supercomputer to probe the endless sea of pi.

Before the Chudnovsky brothers built m zero, Gregory had to derive pi over the Internet while lying in bed. It was inconvenient. The work typically went like this:

Tapping at a small wireless keyboard, which he places on the blankets of his bed, he stares at a computer display screen on one of the bookshelves beside his bed.

The keyboard and screen are connected through cyberspace into the heart of a Cray supercomputer at the Minnesota Supercomputer Center, in Minneapolis. He calls up the Cray through the Internet. When the Cray answers, he sends into the Cray a little software program that he has written. This program—just a few lines of code—tells the supercomputer to start making an approximation of pi. The job begins to run. The Cray starts trying to estimate the number of times the diameter of a circle goes around the periphery.

While this is happening, Gregory sits back on his pillows and waits. He watches messages from the Cray flow across his display screen. The supercomputer is estimating pi. He gets hungry and wanders into the dining room to eat dinner with his wife and his mother. An hour or so later, back in bed, he takes up a legal pad and a red felt-tip pen and plays around with number theory, trying to discover hidden properties of numbers. All the while, the Cray in Minneapolis has been trying to get closer to pi at a rate of a hundred million operations per second. Midnight arrives. Gregory dozes beside his computer screen. Once in a while, he taps on the keys, asking the Cray how things are going. The Cray replies that the job is still active. The night passes and dawn comes near, and the Cray is still running deep toward pi. Unfortunately, since the exact ratio of the circle's circumference to its diameter dwells at infinity, the Cray has not even begun to pinpoint pi. Abruptly, a message appears on Gregory's screen: LINE IS DISCONNECTED.

"What's going on?" Gregory exclaims.

Moments later, his telephone rings. It's a guy in Minneapolis who's working the night shift as the system operator of the Cray. He's furious. "What the hell did you do? You've crashed the Cray! We're down!"

Once again, pi has demonstrated its ability to give the most powerful computers a heart attack.

PI WAS BY NO MEANS the only unexplored number in the Chudnovskys' inventory, but it was one that interested them. They wondered whether the digits contained a hidden rule, an as yet unseen architecture, close to the mind of God. A subtle and fantastic order might appear in the digits of pi way out there somewhere; no one knew. No one had ever proved, for example, that pi did not turn into a string of nines and zeros, spattered in some peculiar arrangement. It could be any sort of arrangement, just so long as it didn't repeat periodically; for it has been proven that pi never repeats periodically. Pi could, however, conceivably start doing something like this: 122333444455555666666. . . . That is, the digits might suddenly shift into a strong pattern. Such a pattern is very regular, but it doesn't repeat periodically. (Mathematicians felt it was very unlikely that pi would ever become obviously regular in some way, but no one had been able to prove that it *didn't.*)

If we were to explore the digits of pi far enough, they might resolve into a breathtaking numerical pattern, as knotty as *The Book of Kells,* and it might mean something. It might be a small but interesting message from God, hidden in the crypt of the circle, awaiting notice by a mathematician. On the other hand, the digits of pi might ramble forever in a hideous cacophony, which was a kind of absolute perfection to a mathematician like Gregory Chudnovsky. Pi looked "monstrous" to him. "We know absolutely *nothing* about pi," he declared from his bed. "What the hell does it mean? The definition of pi is really very simple—it's just the circumference to the diameter—but the complexity of the sequence it spits out in digits is really unbelievable. We have a sequence of digits that looks like gibberish."

"Maybe in the eyes of God pi looks perfect," David said, standing in a corner of the bedroom, his head and shoulders visible above towers of paper.

Mathematicians call pi a transcendental number. In simple terms, a transcendental number is a number that exists but can't be expressed in any finite series of finite operations.* For example, if you try to express pi as the solution to an algebraic equation made up of terms that have integer coefficients in them, you will find that the equation goes on forever. Expressed in digits, pi extends into the distance as far as the

*More precisely: a transcendental number cannot be expressed as the exact solution to any polynomial equation that has a finite number of terms with integer coefficients.

eye can see, and the digits don't repeat periodically, as do the digits of a rational number. Pi slips away from all rational methods used to locate it. Pi is a transcendental number because it transcends the power of algebra to display it in its totality.

It turns out that *almost all* numbers are transcendental, yet *only a tiny handful* of them have ever actually been discovered by humans. In other words, humans don't know anything about *almost all numbers.* There are certainly vast classes and categories of transcendental numbers that have never even been conjectured by humans—we can't even imagine them. In fact, it's very difficult even to prove that a number is transcendental. For a while, mathematicians strongly suspected that pi was a transcendental number, but they couldn't prove it. Eventually, in 1882, a German mathematician named Ferdinand von Lindemann proved the transcendence of pi. He proved, in effect, that pi can't be written on any piece of paper, no matter how big: a piece of paper as big as the universe would not even begin to be large enough to hold the tiniest droplet of pi. In a manner of speaking, pi is undescribable and cannot be found.

The earliest known reference to pi in human history occurs in a Middle Kingdom papyrus scroll, written around 1650 B.C.E. by a scribe named Ahmes. He titled his scroll "The Entrance into the Knowledge of All Existing Things." He led his readers through various mathematical problems and solutions, and toward the end of the scroll he found the area of a circle, using a rough sort of pi.

Around 200 B.C.E., Archimedes of Syracuse found that pi is somewhere between $3\frac{10}{71}$ and $3\frac{1}{7}$. That's about 3.14. (The Greeks didn't use decimals.) Archimedes had no special term for pi, calling it "the perimeter to the diameter." By in effect approximating pi to two places after the decimal point, Archimedes narrowed down the suspected location of pi to one part in a hundred. After that, knowledge of pi bogged down. Finally, in the seventeenth century, a German mathematician named Ludolph van Ceulen approximated pi to thirty-five decimal places, or one part in a hundred million billion billion billion—a calculation that took Ludolph most of his life to accomplish. It gave him such satisfaction that he had the thirty-five digits of pi engraved on his tombstone, which ended up being installed in a special graveyard for professors in St. Peter's Church in Leiden, in the Netherlands. Ludolph was so admired for his digits that pi came to be called the Ludolphian number. But then his tombstone vanished from the graveyard, and some people think it was turned into a sidewalk slab. If so, somewhere in Leiden people are probably walking over Ludolph's digits. The Germans still call pi the Ludolphian number.

In the eighteenth century, Leonhard Euler, mathematician to Catherine the Great, empress of Russia, began calling it *p* or *c*. The first person to use the Greek letter π was William Jones, an English mathematician, who coined it in 1706. Jones probably meant π to stand for "periphery."

It is hard to ignore the ubiquity of pi in nature. Pi is obvious in the disks of the moon and the sun. The double helix of DNA revolves around pi. Pi hides in the rainbow and sits in the pupil of the eye, and when a raindrop falls into water, pi emerges in the spreading rings. Pi can be found in waves and spectra of all kinds, and therefore pi occurs in colors and music, in earthquakes, in surf. Pi is everywhere in superstrings, the hypothetical loops of energy that may vibrate in many dimensions, forming the essence of matter. Pi occurs naturally in tables of death, in what is known as a Gaussian distribution of deaths in a population. That is, when a person dies, the event "feels" the Ludolphian number.

It is one of the great mysteries why nature seems to know mathematics. No one can suggest why this should be so. Eugene Wigner, the physicist, once said that the miracle in the way the language of mathematics fits the laws of physics "is a wonderful gift which we neither understand nor deserve." We may not understand or deserve pi, but nature is aware of it, as Captain O. C. Fox learned while he was recovering in a hospital from a wound that he got in the American Civil War. Having nothing better to do with his time than lie in bed and derive pi, Captain Fox spent a few weeks tossing pieces of fine steel wire onto a wooden board ruled with parallel lines. The wires fell randomly across the lines in such a way that pi emerged in the statistics. After throwing his wires on the floor eleven hundred times, Captain Fox was able to derive pi to two places after the decimal point—he got it to the same accuracy that Archimedes did. But Captain Fox's method was not efficient. Each digit took far more time to get than the previous one. If he had had a thousand years to recover from his wound, he might have gotten pi to perhaps another decimal place. To go deeper into pi, it is necessary to use a machine.

The race toward pi happened in cyberspace, inside supercomputers. In the beginning, computer scientists used pi as an ultimate test of a machine. Pi is to a computer what the East Africa rally is to a car. In 1949, George Reitwiesner, at the Ballistic Research Laboratory, in Maryland, derived pi to 2,037 decimal places with the ENIAC, the first general-purpose electronic digital computer. Working at the same laboratory, John von Neumann (one of the inventors of the ENIAC), searched those digits for signs of order but found nothing he could put his finger on. A decade later, Daniel Shanks and John W. Wrench, Jr., approximated pi to a hundred thousand decimal places with an IBM 7090 mainframe computer, and saw nothing. This was the Shanks-Wrench pi, a milestone. The race continued in a desultory fashion. Eventually, in 1981, Yasumasa Kanada, the head of a team of computer scientists at Tokyo University, used an NEC supercomputer, a Japanese machine, to compute two million digits of pi. People were astonished that anyone would bother to do it, but that was only the beginning of the affair. In 1984, Kanada and his team got sixteen million digits of pi. They noticed nothing remarkable. A year later, William Gosper, a mathematician and distinguished hacker employed at Symbolics, Inc., in Sunnyvale, California, computed pi to seventeen and a half million places with a smallish workstation, beating Kanada's team by a million-and-a-half digits. Gosper saw nothing of interest.

The next year, David H. Bailey, at NASA, used a Cray supercomputer and a formula discovered by two brothers, Jonathan and Peter Borwein, to scoop twenty-nine million digits of pi. Bailey found nothing unusual. A year after that, Kanada and his Tokyo team got 134 million digits of pi. They saw no patterns anywhere. Kanada stayed in to the game. He went past two hundred million digits, and saw further amounts of nothing. Then the Chudnovsky brothers (who had not previously been known to have any interest in calculating pi) suddenly announced that they had obtained 480 million digits of pi—a world record—using supercomputers at two sites in the United States.

Kanada's Tokyo team seemed to be taken by surprise. The emergence of the Chudnovskys as competitors sharpened the Tokyo team's appetite for more pi. They got on a Hitachi supercomputer and ripped through 536 million digits of pi, beating the Chudnovsky brothers and setting a new world record. They saw nothing new in pi. The brothers responded by smashing through *one billion* digits. Kanada's restless boys and their Hitachi were determined not to be beaten, and they soon pushed into *slightly more* than a billion digits. The Chudnovskys took up the challenge and squeaked past the Japanese team again, having computed pi to 1,130,160,664 decimal places, without finding anything special. It was another world record. At this point, the brothers gave up, out of boredom.

If a billion decimals of pi were printed in ordinary type, they would stretch from New York City to the middle of Kansas. This notion raises a question: What is the point of computing pi from New York to Kansas? That question was indeed asked among mathematicians, since an expansion of pi to only forty-seven decimal places would be sufficiently precise to inscribe a circle around the visible universe that doesn't deviate from perfect circularity by more than the distance across a single proton. A billion decimals of pi go so far beyond that kind of precision, into such a lunacy of exactitude, that physicists will never need to use the quantity in any experiment—at least, not for any physics we know of today. The mere thought of a billion decimals of pi gave some mathematicians a feeling of indefinable horror, and they declared the Chudnoskys' effort trivial.

I asked Gregory if an impression I had of mathematicians was true, that they spend a certain amount of time declaring one another's work trivial. "It is true," he admitted. "There is actually a reason for this. Because once you know the solution to a problem it usually is trivial."

For that final, record-breaking, Hitachi-beating, transbillion-digit push into pi, Gregory did the calculation from his bed in New York, working on the Internet with the Cray supercomputer in Minneapolis, occasionally answering the phone when the system operator called to ask why the Cray had crashed. Gregory also did some of the pi work on a massive IBM dreadnought mainframe at the Thomas J. Watson Research Center, in Yorktown Heights, New York, where he also triggered some dramatic crashes. The calculation of more than a billion digits of pi took half a year. This was because the Chudnovsky brothers could get time on the supercomputers only in bits and pieces, usually during holidays and in the dead of night.

Meanwhile, supercomputer system operators had become leery of Gregory. They worried that he might really toast a $30 million supercomputer. The work of calculating pi was also very expensive for the Chudnovskys. They had to rent time on the Cray. This cost the Chudnovskys $750 an hour. At that rate, a single night of driving the Cray into pi could easily cost the Chudnovskys close to ten thousand dollars. The money came from the National Science Foundation. Eventually the brothers concluded that it would be cheaper to build their own supercomputer in Gregory's apartment. They could crash their machine all they wanted in privacy at home, while they opened doors in the house of numbers.

When I first met them, the brothers had got an idea that they would compute pi to two billion digits with their new machine. They would try to almost double their old world record and leave the Japanese team and their sleek Hitachi burning in a gulch, as it were. They thought that testing their new supercomputer with a massive amount of pi would put a terrible strain on their machine. If the machine survived, it would prove its worth and power. Provided the machine didn't strangle on digits, they planned to search the huge resulting string of pi for signs of hidden order. In the end, if what the Chudnovsky brothers ended up seeing in pi was a message from God, the brothers weren't sure what God was trying to say.

GREGORY SAID, "Our knowledge of pi was barely in the millions of digits—"

"We need many billions of digits," David said. "Even a billion digits is a drop in the bucket. Would you like a Coca-Cola?" He went into the kitchen, and there was a horrible crash. "Never mind, I broke a glass," he called. "Look, it's not a problem." He came out of the kitchen carrying a glass of Coca-Cola on a tray, with a paper napkin under the glass, and as he handed it to me he urged me to hold it tightly, because a Coca-Cola spilled into— He didn't want to think about it; it would set back the project by months. He said, "Galileo had to build his telescope—"

"Because he couldn't afford the Dutch model," Gregory said.

"And we have to build our machine, because we have—"

"No money," Gregory said. "When people let us use their supercomputer, it's always done as a kindness." He grinned and pinched his finger and thumb together. "They say, 'You can use it as long as nobody *complains.*' "

I asked the brothers when they planned to build their supercomputer.

They burst out laughing. "You are sitting inside it!" David roared.

"Tell us how a supercomputer should look," Gregory said.

I started to describe a Cray to the brothers.

David turned to his brother and said, "The interviewer answers our questions. It's Pirandello! The interviewer becomes a person in the story." David turned to me and said, "The problem is, you should change your thinking. If I were to put inside this Cray a chopped-meat machine, you wouldn't know it was a meat chopper."

"Unless you saw chopped meat coming out of it. Then you'd suspect it wasn't a Cray," Gregory said, and the brothers cackled.

"In a few years, a Cray will fit in your pocket," David said.

Supercomputers are evolving incredibly fast. The definition of a supercomputer is simply this: one of the fastest and most powerful computers in the world, for its time. M zero was not the only ultrapowerful silicon engine to gleam in the Chudnovskys' designs. They had fielded a supercomputer named Little Fermat, which they had designed with Monty Denneau, a supercomputer architect at IBM, and Saed Younis, a graduate student at the Massachusetts Institute of Technology. Little Fermat was seven feet tall. It sat in a lab at MIT, where it considered numbers.

What m zero consisted of was a group of high-speed processors linked by cables (which covered the floor of the room). The cables formed a network among the processors that the Chudnovskys called a web. On a piece of paper, Gregory sketched the layout of the machine.

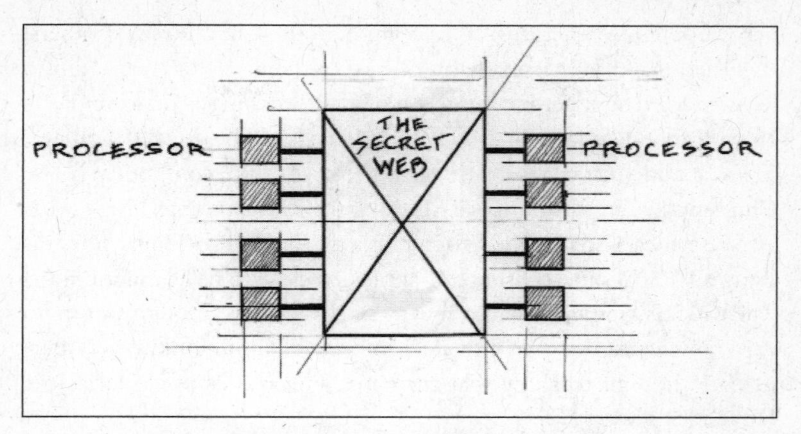

The design of the supercomputer m zero.
Drawing by Richard Preston

He drew a box and put an X through it, to show the web, or network. Then he attached some processors to the web.

The exact design of this web was a secret. "Each processor is connected to all the others," Gregory said. "It's like a telephone network—everybody is talking to everybody else." This made the machine very fast. They planned to have 256 processors. "We will be able to fit them into the apartment," Gregory said. The brothers wrote the machine's software in FORTRAN, a programming language that is "a dinosaur from the late fifties," Gregory said, adding, "There is always new life in this dinosaur." He said that it was very hard to know what exactly was happening inside the machine when it was running. It seemed to have a life of its own.

The brothers would not disclose the exact shape of the network inside their machine. The design contained several new discoveries in number theory, which the Chudnovskys hadn't published. They claimed that they needed to protect their competitive edge in the worldwide race to develop ultrafast computers. "Anyone with a hundred million dollars and brains could be our competitor," David said dryly.

One day, I called Paul Messina, a Caltech scientist and leading supercomputer designer, to get his opinion of the Chudnovsky brothers. It turned out that Messina hadn't heard of them. As for their claim to have built a true supercomputer out of mail-order parts for around seventy thousand dollars, he flatly believed it. "It can be done, definitely," Messina said. "Of course, that's just the cost of the components. The Chudnovskys are counting very little of their human time."

Yasumasa Kanada, the brothers' pi rival at Tokyo University, was using a Hitachi supercomputer that burned close to half a million watts when it was running—half a megawatt, practically enough power to drive an electric furnace in a steel mill. The Chudnovskys were particularly hoping to show that their machine was as powerful as the Hitachi.

"Pi is the best stress test for a computer," David said.

"We also want to find out what makes pi different from other numbers. Eh, it's a business," Gregory said.

David pulled his Mini Maglite flashlight out of his pocket and shone it into a bookshelf, rooted through some file folders, and handed me a color photograph of pi. "This is a pi-scape," he said.

The photograph showed a mountain range in cyberspace: bony peaks and ridges cut by valleys. The mountains and valleys were splashed with colors—yellow, green, orange, violet, and blue. It was the first eight million digits of pi, mapped as a fractal landscape by an IBM supercomputer at Yorktown Heights, which Gregory had programmed from his bed. Apart from its vivid colors, pi looks like the Himalayas.

Gregory thought that the mountains of pi seemed to contain, possibly, a hidden structure. "I see something systematic in this landscape," he said. "It may be just an attempt by the brain to translate some random visual pattern into order." But as he gazed into the nature beyond nature, he wondered if he stood close to a revelation about the circle and its diameter. "Any very high hill in this picture, or any flat plateau, or deep valley would be a sign of *something* in pi," he said. "There seem to be, perhaps, slight variations from randomness in this landscape. There are, perhaps, fewer peaks and valleys than you would expect if pi were truly random, and the peaks and valleys tend to stay high or low a little longer than you'd expect." In a manner of speaking, the mountains of pi looked to him as if they'd been molded by the hand of the Nameless One, *Deus absconditus* (the hidden God). Yet he couldn't really express in words what he thought he saw. To his great frustration, he couldn't express it in the language of mathematics, either. "Exploring pi is like exploring the universe," David remarked.

"It's more like exploring underwater," Gregory said. "You are in the mud, and everything looks the same. You need a flashlight to see anything. Our computer is a flashlight."

David said, "Gregory—I think, really—you are getting tired."

A fax machine in a corner beeped and emitted paper. It was a message from a hardware dealer in Atlanta. David tore off the paper and stared at it. "They didn't ship it! I'm going to kill them! This is a service economy. Of course, you know what that means—the service is terrible."

"We collect price quotes by fax," Gregory said.

"It's a horrible thing. Window-shopping in computerland. We can't buy everything—"

"Because everything won't *exist*," Gregory broke in, and cackled.

"We only want to build a machine to compute a few transcendental numbers—"

"Because we are not licensed for transcendental meditation," Gregory said.

"Look, we are getting nutty," David said.

"We are not the only ones," Gregory said. "We are getting an average of one letter a month from someone or other who is trying to prove Fermat's Last Theorem."

I asked the brothers if they had published any of their digits of pi in a book.

Gregory said that he didn't know how many trees you would have to grind up to publish a billion digits of pi in a book. The brothers' pi had been published on fifteen hundred microfiche cards stored somewhere in Gregory's apartment. The cards held three hundred thousand pages of data, a slug of information much bigger than the *Encyclopaedia Britannica* and containing but one entry, "Pi." David offered to find the cards for me. They had to be around here somewhere. He switched on the lights in the hallway and began rifling through boxes. Gregory got up and began fishing through bookshelves.

"Please sit down, Gregory," David said. Finally the brothers confessed that they had temporarily lost their billion digits of pi. "Look, it's not a problem," David said. "We keep it in different places." He reached inside m zero and pulled out a metal box. It was a naked hard drive, studded with chips. He handed me the object. It hummed gently. "There's pi stored on it. You are holding some pi in your hand."

MONTHS PASSED before I visited the Chudnovskys again. They had been tinkering with their machine and getting it ready to go after two billion digits of pi when Gregory developed an abnormality related to one of his kidneys. He went to the hospital and had some CAT scans made of his torso, to see what things looked like in there. The brothers were disappointed in the quality of the pictures, and they persuaded the doctors to give them the CAT scan data. They processed it in m zero and got detailed color images of Gregory's insides, far more detailed than any image from a CAT scanner. Gregory wrote the imaging software; it took him a few weeks. "There's a lot of interesting mathematics in the problem of making an image of a body," he remarked. It delayed the brothers' probe into the Ludolphian number.

Spring arrived, and Federal Express was active at the Chudnovskys' building, while the superintendent remained in the dark about what was going on. The brothers began to calculate pi. Slowly at first, then faster and faster. In May, the weather warmed up and Con Edison betrayed the brothers. A heat wave caused a brownout in New York City, and as it struck, m zero automatically shut down and died. Afterward, the brothers couldn't get electricity running properly through the machine. They spent two weeks restarting it, piece by piece.

Then, on Memorial Day weekend, as the calculation was beginning to progress, Malka Benjaminovna suffered a heart attack. Gregory was alone with his mother in the apartment. He gave her chest compressions and breathed air into her lungs, although later David couldn't understand how his brother hadn't killed himself saving her.

An ambulance rushed her to St. Luke's-Roosevelt Hospital. The brothers were terrified that they would lose her, and the strain almost killed David. One day, he fainted in his mother's hospital room and threw up blood. He had developed a bleeding ulcer. "Look, it's not a problem," he said to me. After Malka Benjaminovna had been moved out of intensive care, Gregory rented a laptop computer, plugged it into a telephone line in her hospital room, and talked to m zero over the Internet, driving his supercomputer toward pi and watching his mother's blood pressure at the same time.

Malka Benjaminovna improved slowly. When she got home from the hospital, the brothers settled her back in her room in Gregory's apartment and hired a nurse to look after her. I visited them shortly after that, on a hot day in early summer. David answered the door. There were blue half circles under his eyes, and he had lost weight. He smiled weakly and greeted me by saying, "I believe it was Oliver Heaviside, the English physicist, who once said, 'In order to know soup, it is not necessary to climb into a pot and be boiled.' But look, my dear fellow, if you want to be boiled you are welcome to come in." He led me down the dark hallway. Malka Benjaminovna was asleep in her bedroom, and the nurse was sitting beside her. Her room was lined with her late husband Volf's bookshelves, and they were packed with paper. It was an overflow repository.

"Theoretically, the best way to cool a supercomputer is to submerge it in water," Gregory said, from his bed in the junkyard.

"Then we could add goldfish," David said.

"That would solve all our problems."

"We are not good plumbers, Gregory. As long as I am alive, we will not cool a machine with water."

"What is the temperature in there?" Gregory asked, nodding toward m zero's room.

"It grows to above ninety Fahrenheit. This is not good. Things begin to fry."

David took Gregory under the arm, and we passed through the French door into gloom and pestilential heat. The shades were drawn, the lights were off, and an air conditioner in a window ran in vain. Sweat immediately began to pour down my body. "I don't like to go into this room," Gregory said. The steel frame in the center of the room—the heart of m zero—seemed to have acquired more guts, and red lights blinked inside it. I could hear disk drives murmuring. The drives were copying and recopying huge segments of transcendental numbers, to check the digits for perfect accuracy. "If the machine makes an error in a single digit of pi, then every digit after that is nonsense. What comes out is not pi at all, it's just some random number." Thus they had to keep checking and rechecking the digits to make sure they were exactly pi to the last digit.

Gregory knelt on the floor, facing the steel frame.

David opened a cardboard box and removed an electronic board. He began to fit it into m zero. I noticed that his hands were marked with small cuts, which he had got from reaching into the machine.

"David, could you give me the flashlight?" Gregory asked.

David pulled the Mini Maglite from his shirt pocket and handed it to Gregory. The brothers knelt beside each other, Gregory shining the flashlight into the supercomputer. David reached inside with his fingers and palpated a logic board.

"Don't!" Gregory said. "Okay, look. No! No!" They muttered to each other in Russian. "It's too small," Gregory said.

David adjusted an electric fan. "We bought it at a hardware store down the street," he said to me. "We buy our fans in the winter. It saves money." He pointed to a gauge that had a dial on it. "Here we have a meat thermometer."

The brothers had thrust the meat thermometer between two circuit boards inside m zero, in order to look for hot spots. The thermometer's dial was marked "Beef Rare—Ham—Beef Med—Pork."

"You want to keep the machine below 'Pork,' " Gregory remarked.

He lifted a keyboard out of a steel frame and typed something on it. Numbers filled a display screen. "The machine is checking its memory," he said. A buzzer sounded. "It shut down!" he said. "It's a disk-drive controller. The stupid thing obviously has problems."

"It's mentally deficient," David commented. He went over to a bookshelf and picked up a hunting knife. I thought he was going to plunge it into the supercomputer, but he used it to rip open a cardboard box. "We're going to ship the part back to the manufacturer," he said to me. "You had better send it in the original box or you may not get your money back. Now you know the reason this apartment is full of empty boxes. Gregory, I wonder if you are tired."

"If I stand up now, I will fall down," Gregory said. "Therefore, I will sit in my center of gravity. Let me see, meanwhile, what is happening with this machine." He typed something on his keyboard. "You won't believe it, Dave, but the controller now seems to work."

"We need to buy a new one anyway," David said.

"Try Nevada."

David dialed a computer-parts wholesaler in Nevada called Searchlight Compugear. He said loudly, in a Russian accent, "Hi, Searchlight. I need a fifteen-forty controller. . . . No! No! No! I don't need anything else! Just a naked unit! How much you charge? What? Two hundred and fifty-seven dollars . . . ?!"

Gregory glanced at his brother and shrugged. "Eh."

"Look, Searchlight, can you ship it to me Federal Express? For tomorrow morning. How much? . . . *Thirty-nine dollars* for FedEx? Come on! What about afternoon delivery? . . . *Twenty*-nine dollars before three P.M.? *Relax.* What is your name? . . . Bob. Fine. Okay. So, Bob, it is two hundred and fifty-seven dollars plus twenty-nine dollars for FedEx?"

"Twenty-nine dollars for Fed Ex!" Gregory burst out. "It should be fifteen." He pulled a second keyboard out of the frame and tapped the keys. Another display screen came alive and filled with numbers.

"Tell me this," David said to Bob in Nevada. "Do you have a thirty-day money-back guarantee? . . . No? Come on! Look, any device might not work."

"Of course, a part *might* work," Gregory muttered to his brother. "But usually it doesn't."

"Question Number Two: The FedEx should not cost twenty-nine bucks," David said to Bob. "No, nothing! I'm just asking." David hung up the phone. "I'm going to A.K.," he said. "Hi, A.K., this is David Chudnovsky calling from New York. A.K., I need another controller, like the one you sent. Can you send it today FedEx? . . . How much you charge? . . . Naked! I want a naked unit! Not in a shoe box, nothing!"

A rhythmic clicking sound came from one of the disk drives. Gregory remarked to me, "We are calculating pi right now."

"Do you want my MasterCard? Look, it's really imperative that I get my unit tomorrow. Please, I really need my unit bad." David hung up the telephone and sighed. "This is what has happened to a pure mathematician."

"Gregory and David are both extremely childlike, but I don't mean childish at all," Gregory's wife, Christine Pardo Chudnovsky, said one muggy summer day, at the dining room table. "There is a certain amount of play in everything they do, a certain amount of fooling around between two brothers." She was six years younger than Gregory; she had been an undergraduate at Barnard College when she first met him. "I fell in love with Gregory immediately. His illness came with the package." She remained in love with him, even if at times they fought over the heaps of paper. ("I don't have room to put my things down anywhere," she told him.) As we talked, though, pyramids of boxes and stacks of paper leaned against the dining room windows, pressing against the glass and blocking daylight, and a smell of hot electrical gear crept through the room. "This house is an example of mathematics in family life," she said. At night, she dreamed that she was dancing from room to room through an empty apartment that had parquet floors.

David brought his mother out of her bedroom, settled her at the table, and kissed her on the cheek. Malka Benjaminovna seemed frail but alert. She was a small, white-haired woman with a fresh face and clear blue eyes who spoke limited English. A mathematician once described Malka Benjaminovna as the glue that held the Chudnovsky family together. She'd been an engineer during the Second World War, when she designed buildings, laboratories, and proving grounds for testing the Katyusha rocket. Later, she taught engineering at schools around Kiev. Smiling, she handed me plates of roast chicken, kasha, pickles, cream cheese, brown bread, and little wedges of Laughing

Cow cheese in foil. "Mother thinks you aren't getting enough to eat," Christine said. Malka Benjaminovna slid a jug of Gatorade across the table at me.

After we finished lunch, and were fortified with Gatorade, the brothers and I went into the chamber of m zero, into a pool of thick heat. The room enveloped us like noon on the Amazon, and it teemed with hidden activity. The machine clicked, the red lights flashed, the air conditioner hummed, and you could hear dozens of whispering fans. Gregory leaned on his cane and stared into the machine. "Frankly, I don't know what it's doing right now. It's doing some algebra, and I think it's also backing up some pieces of pi."

"Sit down, Gregory, or you will fall," David said.

"What is it doing now, Dave?"

"It's blinking."

"It will die soon."

"Gregory, I heard a funny noise."

"You really heard it? Oh, God, it's going to be like the last time."

"That's it!"

"We are dead! It crashed!"

"Sit down, Gregory, for God's sake!"

Gregory sat on a stool and tugged at his beard. "What was I doing before the system crashed? With God's help, I will remember." He jotted a few notes in a notebook. David slashed open a cardboard box with his hunting knife and lifted something out of the box and plugged it into m zero. Gregory crawled under a table. "Oh, crap," he said from beneath the table.

"Gregory! You killed the system again!"

"Dave, Dave, can you get me a flashlight?"

David handed his Mini Maglite under the table. Gregory joined some cables together and stood up. "Whoo! Very uncomfortable. David, boot it up."

"Sit down for a moment."

Gregory slumped in a chair.

"This monster is going on the blink," David said, tapping a keyboard.

"It will be all right."

On a screen, m zero declared, "The system is ready."

"Ah," David said.

The machine began to click, while its processors silently multiplied and joined huge numbers, heading ever deeper into pi. Gregory went off to bed, David holding him by the arm.

In his junkyard, his nest, his chamber of memory and imagination, Gregory kicked off his gentleman's slippers, lay down on his bed, and brought into his mind's eye the shapes of computing machines yet unbuilt.

IN THE NINETEENTH CENTURY, mathematicians attacked pi with the help of human computers. The most powerful of these was a man named Johann Martin Zacharias Dase. A prodigy from Hamburg, Dase could multiply large numbers in his head. He made a living exhibiting himself to crowds and hiring himself out as a computer for use by mathematicians. A mathematician once asked Dase to multiply 79,532,853 by 93,758,479, and Dase gave the right answer in fifty-four seconds. Dase extracted the square root of a hundred-digit number in fifty-two minutes, and he was able to multiply a couple of hundred-digit numbers in his head in slightly less than nine hours. Dase could do this kind of thing for weeks on end, even as he went about his daily business. He would break off a calculation at bedtime, store everything in his memory for the night, and resume the calculation in the morning. Occasionally, Dase had a system crash. In 1845, he crashed while he was trying to demonstrate his powers to an astronomer named Heinrich Christian Schumacher—he got wrong every multiplication he tried. He explained to Schumacher that he had a headache. Schumacher also noticed that Dase did not in the least *understand* mathematics. A mathematician named Julius Petersen once tried in vain for six weeks to teach Dase the rudiments of geometry—such things as an equilateral triangle and a circle—but they absolutely baffled Dase. He had no problem with large numbers. In 1844, L. K. Schulz von Strassnitsky hired him to compute pi. Dase ran the job in his brain for almost two months. At the end of that time he wrote down pi correctly to the first two hundred decimal places—then a world record.

To many mathematicians, mathematical objects such as a circle seem to exist in an external, objective reality. Numbers, as well, seem to have a reality apart from time or the world. Numbers seem to transcend the universe. Numbers might even exist if the universe did not. I suspect that in their hearts most working mathematicians are Platonists, in that they accept the notion that mathematical reality stands apart from the world, and is at least as real as the world, and possibly gives shape to the world, as Plato suggested. Most mathematicians would probably agree that the ratio of the circle to its diameter exists luminously and eternally in the nature beyond nature, and would exist even if the human mind was not aware of it. Pi might exist even if God had not bothered to create it. One could imagine that pi existed before the universe came into being and will exist after the universe is gone. Pi may even exist apart from God. This is in the opinion of some mathematicians, anyway, who would argue that while there is at least some reason to doubt the existence of God, there is no good reason to doubt the existence of the circle.

"To an extent, pi is more real than the machine that is computing it," Gregory remarked to me one day. "Plato was right. I am a Platonist. Since pi is there, it exists. What we are doing is really close to experimental physics—we are 'observing pi.' Observing pi is easier than studying physical phenomena, because you can prove things in mathematics, whereas you can't prove anything in physics. And, unfortunately, the laws of physics change once every generation."

"Is mathematics a form of art?" I asked.

"Mathematics is partially an art, even though it is a natural science," he said. "Everything in mathematics does exist now. It's a matter of *naming* it. The thing doesn't arrive from God in a fixed form; it's a matter of representing it with symbols. You put it through your mind to make sense of it."

Pi is elusive and can be approached only through approximations. There is no equation built from whole numbers that will give an exact value for pi. If equations are trains threading the landscape of numbers, no train stops at pi. A formula that heads toward pi will never get there, though it can get ever closer to pi. It will consist of a chain of operations that never ends. It is an infinite series. In 1674, Gottfried Wilhelm Leibniz (the coinventor of calculus, along with Isaac Newton) discovered an extraordinary pattern of numbers buried in the circle. This string of numbers—the Leibniz series for pi—has been called one of the most beautiful mathematical discoveries of the seventeenth century:

$$\pi/4 = 1/1 - 1/3 + 1/5 - 1/7 + 1/9 - \ldots$$

In English: pi divided by four equals one minus a third plus a fifth minus a seventh plus a ninth—and so on. It seems almost musical in its harmony. You follow this chain of odd numbers out to infinity, and when you arrive there and sum the terms, you get pi. But since you never arrive at infinity, you never get pi. Mathematicians find it deeply mysterious that a chain of discrete rational numbers can connect so easily to the smooth and continuous circle.

As an experiment in "observing pi," as Gregory put it, I got a pocket calculator and started computing the Leibniz series, to see what would happen. It was easy to do. I got answers that seemed to wander slowly toward pi. As I pushed the buttons on the calculator, the answers touched on 2.66, then 3.46, then 2.89, and 3.34, in that order. The answers landed higher than pi and lower than pi, skipping back and forth across pi, and were gradually closing in on pi. A mathematician would say that the series "converges on pi." It converges on pi forever, playing hopscotch over pi, narrowing it down, but never landing on pi. No matter how far you take it, it never exactly touches pi. Transcendental numbers continue forever, as an endless nonrepeating string, in whatever rational form you choose to display them, whether as digits or an equation. The Leibniz series is a beautiful way to represent pi, and it is finally mysterious, because it doesn't tell us much about pi. Looking at the Leibniz series, you feel the independence of mathematics from human culture. Surely on any world that knows pi the Leibniz series will also be known.

It is worth thinking about what a decimal place means. Each decimal place of pi is a range that shows the *approximate* location of pi to an accuracy ten times as great as the previous range. But as you compute the next decimal place you have no idea where pi will appear in the range. It could pop up in 3, or just as easily in 9, or in 2. The apparent movement of pi as you narrow the range is known as the random walk of pi.

Pi does not move. Pi is a fixed point. The algebra wanders around pi. This is no such thing as a formula that is steady enough and sharp enough to stick a pin into pi. Mathematicians have discovered formulas that converge on pi very fast (that is, they skip around pi with rapidly increasing accuracy), but they do not and cannot hit pi. The Chudnovsky brothers discovered their own formula, a powerful one, and it attacked pi with ferocity and elegance. The Chudnovsky formula for pi was the fastest series for pi ever found that uses rational numbers. It was very fast on a computer. The Chudnovsky formula for pi was thought to be "extremely beautiful" by persons who had a good feel for numbers, and it was based on a torus (a doughnut), rather than on a circle.

The Chudnovsky brothers claimed that the digits of pi form the most nearly perfect random sequence of digits that has ever been discovered. They said that nothing known to humanity appeared to be more deeply unpredictable than the sequence of digits in pi, except, perhaps, the haphazard clicks of a Geiger counter as it detects the decay of radioactive nuclei. But pi isn't random. Not at all. The fact that pi can be produced by a relatively simple formula means that pi is orderly. Pi only *looks* random. In fact, there *has to be a pattern* in the digits. No doubt about it, because pi comes from the most perfectly symmetrical of all mathematical objects, the circle. But the pattern in pi is very, very complex. The Ludolphian number is something fixed in eternity—not a digit out of place, all characters in their proper order, an endless sentence written to the end of the world by the division of the circle's diameter into its circumference.

"Pi is a damned good fake of a random number," Gregory said. "I just wish it were not as good a fake. It would make our lives a lot easier."

Around the three hundred millionth decimal place of pi, the digits go 88888888—eight eights come up in a row. Does this mean anything? It seems to be random noise. Later, ten sixes erupt: 6666666666. Only more noise. Somewhere past the half-billion mark appears the string 123456789. It's an accident, as it were. "We do not have a good, clear, crystallized idea of randomness," Gregory said. "It cannot be that pi is truly random. Actually, a truly random sequence of numbers has not yet been discovered."

He explained that the "random" combinations of a slot machine in a casino are not random at all. They're generated by simple computer programs, and, according to Gregory, the pattern is easy to figure out. "You might need only five consecutive tries on a slot machine to figure out the pattern," he said.

"Why don't you go to Las Vegas and make some money this way?" I asked.

"Eh." Gregory shrugged, leaning on his cane.

"But look, this is not interesting," David said. Besides, he pointed out, Gregory's health would be threatened by a trip to Las Vegas.

No one knew what happened to the digits of pi in the deeper regions, as the number resolved toward infinity. Did the digits turn into nothing but eights and fives, say? Did they show a predominance of sevens? In fact, no one knew if a digit simply stopped appearing in pi. For example, there might be no more fives in pi after a certain point. Almost certainly, pi doesn't do this, Gregory Chudnovsky thinks, because it would be stupid, and nature isn't stupid. Nevertheless, no one has ever been able to prove or disprove it. "We know very little about transcendental numbers," Gregory said.

If you take a string of digits from the square root of two and you compare it to a string of digits from pi, they look the same. There's no way to tell them apart just by looking at the digits. Even so, the two numbers have completely distinct properties. Pi and the square root of two are as different from each other as a Rembrandt is from a Picasso, but human beings don't have the ability to tell the two numbers apart by looking at their digits. (A sufficiently intelligent race of beings could probably do it easily.) Distressingly, the number pi makes the smartest humans into blockheads.

Even if the brothers couldn't prove anything about the digits of pi, they felt that by looking at them through the window of their machine they might have a chance of at least seeing something that could lead to an important conjecture about pi or about transcendental numbers as a class. You can learn a lot about all cats by looking closely at one of them. So if you wanted to look closely at pi, how much of it could you see with a very large supercomputer? What if you turned the entire universe into a computer? What if you took every particle of matter in the universe and used all of it to build a computer? What then? How much pi could you see? Naturally, the brothers had considered this project. They had imagined a supercomputer built from the whole universe.

Here's how they estimated the machine's size. It has been calculated that there may be around 10^{79} electrons and protons in the observable universe. This is the so-called Eddington number of the universe. (Sir Arthur Stanley Eddington, an astrophysicist, first came up with it.) The Eddington number is the digit 1 followed by seventy-nine zeros: 10,000,000,000,000,000,000,000,000,000,000,000,000,000,000,000,000, 000,000,000,000,000,000,000,000,000,000,000. Ten vigintsextillion. The Eddington number. It gives you an idea of the power of the device that the Chudnovskys referred to as the Eddington machine.

The Eddington machine was the entire universe turned into a computer. It was made of all the atoms in the universe. If the Chudnovsky brothers could figure out how to program it with FORTRAN, they might make it churn toward pi.

"In order to study the sequence of pi, you have to store it in the Eddington machine's memory," Gregory said. To be realistic, the brothers felt that a practical Eddington machine wouldn't be able to store more than about 10^{77} digits of pi. That's only a hundredth of the Eddington number. Now, what if the digits of pi were to begin to show regularity only *beyond* 10^{77} digits? Suppose, for example, that pi were only to begin manifesting a regularity starting at 10^{100} decimal places? That number is known as a googol. If the design in pi appeared only after a googol of digits, then not even the largest possible computer would ever be able to penetrate pi far enough to reveal any order in it. Pi would look totally disordered to the universe, even if it contained a slow, vast, delicate structure. A mere googol of pi might be only the first warp and weft, the first knot in a colored thread, in a limitless tapestry woven into gardens of delight and cities and towers and unicorns and unimaginable beasts and impenetrable mazes and unworldly cosmogonies, all invisible forever to us. It may never be possible, in principle, to see the design in the digits of pi. Not even nature itself may know the nature of pi.

"If pi doesn't show systematic behavior until more than ten to the seventy-seven decimal places, it would really be a disaster," Gregory said. "It would actually be horrifying."

"I wouldn't give up," David said. "There might be some way of leaping over the barrier—"

"And of attacking the son of a bitch," Gregory said.

THE BROTHERS first came in contact with the membrane that divides the dreamlike earth from the perfect and beautiful world of mathematical reality when they were boys, growing up in Kiev. Their father, Volf, gave David a book entitled *What is Mathematics?,* written by Richard Courant and Herbert Robbins, two American mathematicians. The book is a classic. Millions of copies of it had been printed in unauthorized Russian and Chinese editions alone. (Robbins wrote most of the book, while Courant got ownership of the copyright and collected most of the royalties but paid almost none of the money to Robbins.) After reading it, David decided to become a mathematician. Gregory soon followed his brother's footsteps into the nature beyond nature.

Gregory's first publication, in a Soviet math journal, came when he was sixteen years old: "Some Results in the Theory of Infinitely Long Expressions." Already you can see where he was headed. David, sensing his younger brother's power, encouraged him to grapple with central problems in mathematics. In 1900, at the dawn of the twentieth century, the German mathematician David Hilbert had proposed a series of twenty-three great problems in mathematics that remained to be solved, and he'd challenged his colleagues, and future generations, to solve them. They became known as the Hilbert problems. At the age of seventeen, Gregory Chudnovsky made his first major discovery when he solved Hilbert's Tenth Problem. To solve a Hilbert problem would be an achievement for a lifetime; Gregory was a high school student who had read a few books on mathematics. Strangely, a young Russian mathematician named Yuri Matyasevich had also just solved Hilbert's Tenth Problem, but Gregory hadn't heard the news. Eventually, Matyasevich said that the Chudnovsky method was the preferred way to solve Hilbert's Tenth Problem.

The brothers enrolled at Kiev State University, and took their PhDs at the Ukrainian Academy of Sciences. At first, they published their papers separately, but as Gregory's health declined, they began collaborating. They lived with their parents in Kiev until the family decided to try to take Gregory abroad for medical treatment. In 1976, Volf and Malka applied to the government of the USSR for permission to emigrate. Volf was immediately fired from his job.

It was a totalitarian state. The KGB began tailing the brothers. "I had twelve KGB agents on my tail," David told me. "No, look, I'm not kidding! They shadowed me around the clock in two cars, six agents in each car—three in the front seat and three in the backseat. That was how the KGB operated." One day in 1976, David was walking down the street when KGB officers attacked him, fracturing his skull. He nearly died. He didn't dare go to the hospital; he went home instead. "If I had gone to the hospital, I would have died for sure," he said. "The hospital was run by the state. I would forget to breathe."

One July day, plainclothesmen from the KGB accosted Volf and Malka on a street corner and beat them up. They broke Malka's arm and fractured her skull. David took his mother to the hospital, where he found that the doctors feared the KGB. "The doctor in the emergency room said there was no fracture," David recalled.

By this time, the Chudnovskys were quite well known to mathematicians in the United States. Edwin Hewitt, a mathematician at the University of Washington, in Seattle, had collaborated with Gregory on a paper. He brought the Chudnovskys' case to the attention of Senator Henry M. "Scoop" Jackson—a powerful politician from Washington State—and Jackson began putting pressure on the Soviets to let the Chudnovsky family leave the country. Not long before that, two members of a French parliamentary delegation made an unofficial visit to Kiev to see what was going on with the Chudnovskys. One of the visitors was Nicole Lannegrace, who would later become David's wife. The Soviet government unexpectedly let the Chudnovskys go. "That summer when I was getting killed by the KGB, I could never have imagined that the next year I would be in Paris in love, or that I would wind up in New York, married to a beautiful Frenchwoman," David said.

IF PI IS TRULY RANDOM, then at times pi will appear to be orderly. Therefore, if pi is random it contains accidental order. For example, somewhere in pi a sequence may run 07070707070707070707 for as many digits as there are atoms in the sun. It's just an accident. Somewhere else the exact same sequence may appear, only this time interrupted, just once, by the digit 3. Another accident. Every possible arrangement of digits probably erupts in pi, though this has never been proved. "Even if pi is not truly random, you can still assume that you get every string of digits in pi," Gregory told me. In this respect, pi is like the Library of Babel in the story by Jorge Luis Borges. In that story, Borges imagined a library of vast size that contained all possible books.

You could find all possible books in pi. If you were to assign letters of the alphabet to combinations of digits—for example, the letter *a* might be 12, the letter *b* might be 34—you could turn the digits of pi into letters. (It doesn't matter what digits are assigned to what letters—the combination could be anything.) You could do this with all alphabets and ideograms in all languages. Then pi could be turned into strings of written words. Then, if you could look far enough into pi, you would probably find the expression "See the U.S.A. in a Chevrolet!" Elsewhere, you would find Christ's Sermon on the Mount in his native Aramaic tongue, and you would find versions of the Sermon on the Mount that are blasphemy. Also, you would find a guide to the pawnshops of Lubbock, Texas. It might or might not be accurate. Even so, somewhere else you *would* find the accurate guide to Lubbock's pawnshops . . . if you could look far enough into pi. You would find, somewhere in pi, the unwritten book about the sea that James Joyce supposedly intended to tackle after he finished *Finnegans Wake*. You would find the collected transcripts of *Saturday Night Live* rendered into Etruscan. You would find a Google-searchable version of the entire Internet with every page on it exactly as it existed at midnight on July 1, 2007, and another version of the Internet from thirty seconds later. Each occurrence of an apparently ordered string in pi, such as the words "Ruin hath taught me thus to ruminate, / That Time will come and take my love away," is followed by unimaginable deserts of babble. No book and none but the shortest poems will ever actually be seen in pi, for it is infinitesimally unlikely that even as brief a text as an English sonnet will appear in the first 10^{77} digits of pi, which is the longest piece of pi that can be calculated in this universe.

Anything that can be produced by a simple method is orderly. Pi can be produced by very simple methods; it is orderly, for sure. Yet the distinction between chance and fixity dissolves in pi. The deep connection between order and disorder, between cacophony and harmony, seems to be tantalizingly almost visible in pi, but not quite. "We are looking for some rules that will distinguish the digits of pi from other numbers," Gregory said. "Think of games for children. If I give you the sequence one, two, three, four, five, can you tell me what the next digit is? A child can do it: the next digit is six. What if I gave you a sequence of a million digits from pi? Could you tell me the next digit just by looking at it? Why does pi look totally unpredictable, with the highest complexity? For all we know, we may never find out the rule in pi."

HERBERT ROBBINS, the coauthor of *What Is Mathematics?*, the book that had turned the Chudnovsky brothers on to math, was an emeritus professor of mathematical statistics at Columbia University and had become friends with the Chudnovskys. He lived in a rectilinear house with a lot of glass in it, in the woods near Princeton, New Jersey. Robbins was a tall, restless man in his seventies, with a loud voice, furrowed cheeks, and penetrating eyes. One day, he stretched himself out on a daybed in a garden room in his house and played with a rubber band, making a harp across his fingertips.

"It is a very difficult philosophical question, the question of what 'random' is," Robbins said. He plucked the rubber band with his thumb, *boink, boink.* "Everyone knows the famous remark of Albert

Einstein, that God does not throw dice. Einstein just would not believe that there is an element of randomness in the construction of the world. The question of whether the universe is a random process or is determined in some way is a basic philosophical question that has nothing to do with mathematics. The question is important. People consider it when they decide what to do with their lives. It concerns religion. It is the question of whether our fate will be revealed or whether we live by blind chance. My God, how many people have been murdered over an answer to that question! Mathematics is a lesser activity than religion in the sense that we've agreed not to kill each other but to discuss things."

Robbins got up from the daybed and sat in an armchair. Then he stood up and paced the room, and sat at a table, and moved himself to a couch, and went back to the table, and finally returned to the daybed. The man was in constant motion.

"Mathematics is broken into tiny specialties today, but Gregory Chudnovsky is a generalist who knows the whole of mathematics as well as anyone," he said as he moved around. "He's like Mozart. I happen to think that his and David's pi project is a will-o'-the-wisp, but what do I know? Gregory seems to be asking questions that can't be answered. To ask for the system in pi is like asking, 'Is there life after death?' When you die, you'll find out. Most mathematicians are not interested in the digits of pi. In order for a mathematician to become interested in a problem, there has to be a possibility of solving it. Gregory likes to do things that are impossible."

The Chudnovsky brothers were operating on their own, and they were looking more and more unemployable. Columbia University was never going to make them full-fledged members of the faculty, never give them tenure. This had become obvious. The John D. and Catherine T. MacArthur Foundation awarded Gregory Chudnovsky a "genius" fellowship. The brothers had won other fashionable and distinguished prizes, but there was a problem in their résumé, which was that Gregory had to lie in bed most of the time. The ugly truth was that Gregory Chudnovsky couldn't get an academic job because he was physically disabled. But there were other, more perplexing reasons that had led the Chudnovskys to pursue their work in solitude. They had been living on modest grants from the National Science Foundation and various other research agencies and, of course, on their wives' salaries. Christine's father, Gonzalo Pardo, who was a professor of dentistry, had also chipped in. He had built the steel frame for m zero in his basement, using a wrench and a hacksaw.

The brothers' solitary mode of existence had become known to mathematicians around the world as the Chudnovsky Problem. Herbert Robbins eventually decided to try to solve it. He was a member of the National Academy of Sciences, and he sent a letter to all of the mathematicians in the academy:

> I fear that unless a decent and honorable position in the American educational research system is found for the brothers soon, a personal and scientific tragedy will take place for which all American mathematicians will share responsibility.

There wasn't much of a response. Robbins got three replies to his letter. One, from a professor of mathematics at an Ivy League university, complained about David Chudnovsky's personality. He remarked that "when David learns to be less overbearing," the brothers might have better luck.

Then Edwin Hewitt, the mathematician who had helped get the Chudnovsky family out of the Soviet Union, got mad, and erupted in a letter to colleagues:

> The Chudnovsky situation is a national disgrace. Everyone says, "Oh, what a crying shame" & then suggests that they be placed at *somebody else's institution.* No one seems to want the admittedly burdensome task of caring for the Chudnovsky family.

The brothers, because they insisted that they were one mathematician divided between two bodies, would have to be hired as a pair. Gregory would refuse to take any job unless David got a job, too, and vice versa. To hire them, a math department would have to create two openings. And Gregory couldn't teach classes in the normal way, because he was more or less confined to bed. And he might die, leaving the Chudnovsky Mathematician bereft of half its brain.

"The Chudnovskys are people the world is not able to cope with, and they are not making it any easier for the world," Herbert Robbins said. "Even so, this vast educational system of ours has poured the Chudnovskys out on the sand, to waste. When I go up to that apartment and sit by Gregory's bed, I think, My God, when I was a mathematics student at Harvard I was in contact with people far less interesting than this. I'm grieving about it."

"TWO BILLION DIGITS OF PI? Where do they keep them?" Samuel Eilenberg said scornfully. Eilenberg was a distinguished topologist and emeritus professor of mathematics at Columbia University.

"I think they store the digits on a hard drive," I answered.

Eilenberg snorted. He didn't care about some spinning piece of metal covered with pi. He was one of the reasons why the Chudnovskys would never get permanent jobs at Columbia; he made it pretty clear that he would see to it that they were denied tenure. "In the academic world, we have to be careful who our colleagues are," he told me. "David is a nudnik! You can spend all your life computing digits. What for? It's about as interesting as going to the beach and counting sand. I wouldn't be caught dead doing that kind of work."

In his view, there was something unclean about doing mathematics with a machine. Samuel Eilenberg was a member of the famous Bourbaki group. This group, a sort of secret society of mathematicians that was founded in 1935, consisted mostly of French members (though Eilenberg was originally Polish) who published collectively under the fictitious name Nicolas Bourbaki; they were referred to as "the Bourbaki." In a quite French way, the Bourbaki were purists who insisted on rigor and logic and formalism. Some members of the Bourbaki group looked down on applied mathematics—that is, they seemed to scorn the use of mathematics to solve real-world problems, even in physics. The Bourbaki especially seemed to dislike the use of machinery in pursuit of truth. Samuel Eilenberg appeared to loathe the Chudnovskys' supercomputer and what they were doing with it. "To calculate the two billionth digit of pi is to me abhorrent," he said.

" 'Abhorrent'? Yes, most mathematicians would probably agree with that," said Dale Brownawell, a respected number theorist at Penn State. "Tastes change, though. To see the Chudnovskys carrying on science at such a high level with such meager support is awe-inspiring."

Richard Askey, a prominent mathematician at the University of Wisconsin at Madison, would occasionally fly to New York to sit at the foot of Gregory Chudnovsky's bed and talk about mathematics. "David Chudnovsky is a very good mathematician," Askey said to me. "Gregory is a great mathematician. The brothers' pi stuff is just a small part of their work. They are really trying to find out what the word 'random' means. I've heard some people say that the brothers are wasting their time with that machine, but Gregory Chudnovsky is a very intelligent man who has his head screwed on straight, and I wouldn't begin to question his priorities. Gregory Chudnovsky's situation is a national problem."

"IT LOOKS LIKE KVETCHING," Gregory said from his bed. "It looks cheap, and it is cheap. I don't think we were somehow wronged. I really can't teach. So what does one do about it? We barely have time to do the things we want to do. What is life, and where does the money come from?" He shrugged.

At the end of the summer, the brothers halted their probe into pi. They had other things they wanted to do with their supercomputer, and it was time to move on. They had surveyed pi to 2,260,321,336 digits. It was a world record, doubling their previous world record. If the digits were printed in type, they would stretch from New York to Los Angeles.

In Japan, their competitor Yasumasa Kanada reacted gracefully. He told *Science News* that he might be able to get a billion and a half digits if he could rent enough time on the Hitachi—the half-megawatt monster.

"You see the advantage of being truly poor," Gregory said to me. "We had to build our machine, but now we own it."

M zero had spent most of its time checking the answer to make sure it was correct. "We have done our tests for patterns, and there is nothing," Gregory said. He was nonchalant about it. "It would be rather stupid if there were a pattern in a few billion digits. There are the usual things. The digit three is repeated nine times in a row, and we didn't see that before. Unfortunately, we still don't have enough computer power to see anything in pi."

And yet . . . and yet . . . the brothers felt that they might have noticed something in pi. It hovered out of reach, but seemed a little closer now. It was a slight change in pi that seemed to rise and fall like a tide, as if a distant moon were passing over the sea of digits. It was something random, probably. The brothers felt that they might only have glimpsed the human desire for order. Or was it a wave rippling through pi? Would the wave, if it was there, be the first thread in a tapestry of worlds blossoming in pi? "We need a trillion digits," David said. Maybe one day they would run the calculation into a trillion digits. Or maybe not. A trillion digits of pi printed in ordinary type would stretch from here to the moon and back, twice. Maybe one day, if they lived and if their machines held together, they would orbit the moon in digits, and would head for Alpha Centauri, seeking pi.

Gregory is lying on his bed in the junkyard, now. He offers to show me the last digits the supercomputer found. He types a command, and suddenly the whole screen fills with pi. It's the raw Ludolphian number, pouring across the screen like Niagara Falls:

72891 51567 97145 46268 92720 56914 19491 70799 30612 27184
95997 75819 61414 47296 81115 92768 25023 87974 42024 32465
81816 25413 12164 96683 83188 86493 16114 55018 80584 26203
71989 99024 98835 10467 22124 63734 94382 70510 64281 32133
84515 75884 47736 80693 93435 69959 13571 88057 62592 60719
58508 38025 73050 11862 43946 99422 06487 07264 08095 58354
41083 43437 83790 00353 73416 69273 76820 40100 54718 28029
00958 45404 09196 25724 40953 10724 75287 88238 71194 22897
36462 82455 69706 19364 35459 84229 95107 39973 54996 68154
14759 50184 95343 60383 37189 76295 12572 70965 58816 94729
09508 25947 06150 01226 73434 26496 86070 41411 62634 95296
69333 80436 51116 81295 92670 33384 07650 40965 11979 85185
50164 21984 40980 27554 25619 05834 95554 34498 43497 55136

88999 51731 69029 01197 60153 45399 73782 80898 99826 36229
28846 77788 04108 11793 89363 51922 14801 13183 14735 68818
49953 27420 48050 19186 07391 11248 22845 78059 61348 96790
18820 54573 01261 27678 17413 87779 66981 15311 24707 34258
41235 99801 92693 52561 92393 53870 24377 10069 16106 22971
02523 30027 49528 06378 64067 12852 77857 42344 28836 88521
72435 85924 57786 36741 32845 66266 96498 68308 59920 06168
63376 85976 35341 52906 04621 44710 52106 99079 33563 54625
71001 37490 77872 43403 57690 01699 82447 20059 93533 82919
46119 87044 02125 12329 11964 10087 41341 42633 88249 48948
31198 27787 03802 08989 05316 75375 43242 20100 43326 74069
33751 86349 40467 52687 79749 68922 29914 46047 47109 31678
05219 48702 00877 32383 87446 91871 49136 90837 88525 51575
35790 83982 20710 59298 41193 81740 92975 31

We observe pi in silence.

A Death in the Forest

In 1911, a woman named Sallie Dooley established a Japanese garden at Maymont, her estate in Richmond, Virginia. She planted bamboo, built a gazebo and a waterfall, and, according to her husband, James Dooley, a financier, "purchased the most costly evergreens from all parts of the world." She died in 1925, leaving Maymont to the city of Richmond. It became a park, and the Japanese garden went untended. In 1951, an entomologist with the Virginia Department of Agriculture discovered a species of Asian insect known as the hemlock woolly adelgid infesting an eastern hemlock—a tree native to North America—on property near Maymont Park. The hemlock woolly adelgid is a tiny brown bug similar to an aphid; the body of an adult is covered with a protective white fluff that makes it look like a fleck of cotton. It is a parasite, and it feeds on several species of hemlock and spruce trees in Asia. This was its first known appearance in eastern North America. The suspicion was that it had come from Sallie Dooley's languishing evergreens, though no one could be sure. Experts considered it a curiosity.

After hatching from an egg, the woolly adelgid goes through a crawler stage, when it moves around. The crawlers are almost invisible to the naked eye. They can drift in the air from tree to tree, and they can cling to the legs and feathers of migrating birds. The insect eventually settles down among the needles of a host tree. It inserts a bundle of mouthparts at the base of a needle and spends the rest of its life—a few months—sucking nutrients out of the tree. A female can lay eggs without being fertilized by a male. The offspring are clones of their mother—genetically identical to her. As it has turned out, the population of woolly adelgids in North America seems to consist entirely of female clones. Males still hatch occasionally, but they breed and live in spruce trees, and American spruces lack certain nutrients they need, so they die—a further indication that the adelgids are transplants. It hardly matters: a single female clone can generate as many as ninety thousand copies of herself in a year.

In Asia, many kinds of natural predators, especially beetles, eat the woolly adelgid, and the host trees have developed resistance to it. In North America, though, there are no natural predators of the adelgid, and eastern hemlocks have virtually no resistance to it. In coming to America, the Asian insect escaped its predators. When millions of woolly adelgids cover the branches of an eastern hemlock, it turns a dirty whitish color, as if it had been flocked with artificial snow. Many of its needles fall off. The tree puts out a new crop of needles the following spring, but the crawlers attach themselves to the new needles, the tree goes into shock, and the needles fall off again. The cycle of shock and defoliation continues until the tree dies, usually in two to six years.

There weren't many eastern hemlocks in Richmond. The tree doesn't occur naturally in the area, but it had been planted in some people's yards, and specimens were scattered sparsely through the city. (Hemlocks are often trimmed into hedges.) For thirty years after its discovery near Maymont Park, the insect gradually moved around the hemlocks of Richmond, and over time many of the the hemlocks in the city lost their needles and died. However, gardeners found that if they sprayed an infested hemlock once a year with pesticides or an oil spray, the bugs would be suppressed.

In the 1980s, an entomologist with the Virginia Department of Forestry named Tim Tigner began tracking the woolly adelgid around Richmond. "We advised people not to worry about it," Tigner said to me recently. "It didn't seem to be doing anything." Then, in the late eighties, Tigner learned that the insect had made its way into a natural stand of ancient hemlocks on the York River, forty miles east of the city, and he went to have a look. He got a shock: 90 percent of the hemlocks were dead. The woolly adelgid had turned the grove into a sun-bleached ruin.

BOTANISTS SOMETIMES REFER to the eastern hemlock as the redwood of the East. It is a tall, long-lived conifer with soft, flat needles and feathery foliage. It has a massive, straight trunk that rises to an impressive height, flaring into a dark, mysterious-looking crown, which is filled with all sorts of living things. The eastern hemlock's species name is *Tsuga canadensis.* It occurs naturally in the Appalachian Mountains from Georgia to New Brunswick and Nova Scotia, with a range that runs westward through Michigan into Wisconsin. The tallest eastern hemlocks are somewhat more than 170 feet high, and the largest ones (measured by volume of wood) can be more than six feet in diameter. The oldest living specimens may be more than six hundred years old. Hemlocks and redwoods are extremely shade tolerant—they can grow in dark places where no other trees can survive. Both kinds of trees do especially well in moist valleys filled with temperate rain forest. It seems that few people know that there are rain forests in California. Possibly even fewer people realize that there are also rain forests in the East.

Hemlocks thrive in the temperate rain forest found in the southern Appalachian Mountains. In simple terms, a temperate rain forest is a cool forest that receives at least 80 inches of rainfall a year. Some parts of the southern Appalachians receive up to 130 inches of rainfall a year, with very little snow—more rain than in many parts of the Amazon basin. In the temperate rain forests of the southern Appalachians, hemlocks grow in moist, cool valleys and on mountain slopes, and they form dense stands in the upland valleys called coves.

Hemlocks cast deep shade, and they cover the ground with beds of needles, altering the temperature, moisture, and chemistry of the soil around them. This creates a distinctive habitat for certain animals and plants. An old-growth forest is a forest that's survived for many centuries without being changed by logging or fire. Only small fragments of old-growth forests remain in the East. Many of them are in Great Smoky Mountains National Park, which lies along the mountainous divide between Tennessee and North Carolina. The national park covers half a million acres; about a fifth of the park has apparently never been logged. Loggers haven't bothered to go into many coves to cut hemlocks, because the tree is practically worthless for lumber: the wood is full of knots, and often fractures when the tree falls. Some ecologists believe that the hemlock coves of the southern Appalachians contain, or until recently contained, the last examples of primeval rain forest in eastern North America—pockets of rain-forest habitat that seem to have remained unchanged for thousands of years.

In 1988, around the time Tim Tigner saw how the woolly adelgid had destroyed a grove of old hemlocks by the York River, the insect was discovered in Shenandoah National Park, in northern Virginia. It seems to have arrived there when crawlers clung to the legs and feathers of migrating birds that visit or nest in hemlock trees—the black-throated green warbler, the solitary vireo. In Shenandoah, the insect got into stands of old hemlocks packed tightly together in coves, and it multiplied with explosive speed. By 1992, most of the hemlocks in the park were infested, and three years later the majority of them were dead. Today, stands of eastern hemlock have essentially disappeared from Shenandoah National Park.

The crawlers spread rapidly northward. They moved southward only slowly, though, possibly because there were few crawlers around in the autumn when birds flew south. By 1998, many of the hemlock groves in the Delaware Water Gap National Recreation Area, which lies between Pennsylvania and New Jersey, were infested and had begun to die. From eastern Pennsylvania to Connecticut, hemlocks were being turned into skeletons. The insect got to Massachusetts. There, stands of old hemlocks were defoliated. However, a spell of intensely cold weather during the winter of 1996, when temperatures in parts of the Northeast fell to as low as twenty degrees below zero, seemed to kill many adelgids. "The hemlocks looked okay after that cold winter," James Akerson, an ecologist with Shenandoah National Park, said. "It may have given us a false sense of hope."

INVASIVE SPECIES OF MICROBES, plants, and animals are changing ecosystems all over the planet in a biological upheaval that may affect almost everything that lives. The cause of the upheaval is the human species. Life on the planet is being homogenized by the expanding human population and the frequent and rapid movement of people and goods, which carry invasive organisms with them. These invasives often flourish in their new ecosystems because, like the woolly adelgid, they have escaped their predators. A fungal disease called chestnut blight, from Asia, first appeared in North America in 1904. Spread by wind, rain, and birds, it killed almost every American chestnut tree. Chestnuts had once saturated vast stretches of forest in the Appalachians. They essentially vanished from the ecosystem. The term biologists use for this is "functional extinction."

Since the 1930s, the American elm has gone almost extinct in the wild, pushed into oblivion by an invasive Asian fungus spread by an invading beetle from Europe. In the 1960s and '70s, the balsam woolly adelgid (from northern Asia) got into the Fraser fir, a native American species growing on the higher ridges and peaks of the southern Appalachian Mountains; this parasite killed from 70 to 90 percent of the mature wild Fraser firs, making the mountains look as if they were covered with driftwood. (Today the wild Fraser firs in the Appalachians often don't get much taller than a person before they die from adelgid infestation.) A fungal disease of unknown origin has killed off the vast majority of the wild flowering dogwoods in North America. Another disease, sudden oak death, has killed hundreds of thousands of oaks in California and may get into Eastern oaks. A European insect carrying a European fungus has lately caused a mass dying of the American beech tree, and the American beech's future as a species in the wild is uncertain. An Asian beetle called the emerald ash borer arrived in Michigan in 2001 in packing wood from China. It is devastating to a number of species of American ash trees. Despite strong efforts to control it, the emerald ash borer keeps appearing in different places, and it seems capable of not only wiping out the ash but threatening the classic major-league baseball bat (which is commonly made of ash). Another invader, the Asian long-horned beetle, had its North American debut in Brooklyn, where it showed up in a park near warehouses that held large amounts of packing wood from China. The Asian long-horned beetle has infested tens of thousands of trees in New Jersey and

Long Island, and it has shown up Sacramento. It could take out the sugar maple. In effect, the trees of North America have been hit with all sorts of Ebolas of their own.

When a parasite moves to a new habitat, it can find new hosts through a process called the trans-species jump. Often the new host has no resistance; the host and the parasite haven't had time to adjust to each other through natural selection. (It is frequently not in the best interest of a parasite to kill its host quickly.) One example is the human immunodeficiency virus, HIV. It appears to have once lived in chimpanzees, though it doesn't make them sick—the chimp's immune system is well acquainted with the virus and has learned how to deal with it. In Africa, at various times and places in the twentieth century, HIV made trans-species jumps into humans—probably through hunters who killed and butchered chimps, and so were exposed to infected chimp blood. Once the virus had escaped the chimpanzee's immune system, it amplified itself freely in its new hosts.

Another example of an invasive species is the Ebola virus. Ebola is a parasite with a known tendency to make trans-species jumps into new hosts. Ebola lives naturally in some unknown type of host in central Africa—possibly a bat, possibly a wingless fly that lives on a bat, or quite possibly some other creature. Ebola probably doesn't make its natural host very sick. Ebola makes primates incredibly sick. This means that primates are *not* the original host of Ebola.

Outbreaks of Ebola in humans tend to burn out fairly quickly, but Ebola is a far more serious matter for gorillas. In recent years, roughly a third of the gorillas in protected areas in central and west Africa have died from Ebola virus. The virus is spreading unchecked in the gorilla population of central and western Africa, and it kills around 90 percent of the gorillas it infects. No one knows how Ebola has been getting into gorillas; possibly some disturbance in the ecosystem has put the animals into contact with the unknown host of Ebola.

I sometimes wonder if the unknown natural host of Ebola is itself an invasive species—some sort of rodent or insect, perhaps—that's moving into disturbed habitats in the African rain forest. If so, this might explain why Ebola seems to be jumping into gorillas more frequently these days. Ebola's host might be moving into new niches that

Hemlock woolly adelgid crawlers.
Artwork by Peter Arkle

have opened up in a rain forest that's being changed by logging and human invasion. The Ebola host might be bringing itself and its parasite—Ebola—into close contact with gorillas. The World Conservation

Union recently put the western gorilla on its critically endangered list; the Ebola virus, together with poaching, could push the western gorilla to extinction in the wild (some gorillas would persist in captivity). In other words, what happened to the American chestnut could also happen to the western gorilla: functional extinction due to a species-jumping parasite.

Global climate change has become entangled with the problem of invasive species. A warmer climate could allow some invaders to spread farther, while causing native organisms to go extinct in their traditional habitats and making room for invaders. The earth's biosphere can be thought of as a sort of palace. The continents are rooms in the palace; islands are smaller rooms. Each room has its own decor and unique inhabitants; many of the rooms have been sealed off for millions of years. Now the doors in the palace have been flung open, and the walls are coming down.

Global climate change may be helping the hemlock adelgids spread both north and south. Winters in the north are becoming steadily warmer, and the insects are not likely to be hit as often with deep cold. Summers in the southern Appalachians have lately become drier and hotter, and drought stress makes infested hemlocks far more susceptible to parasites. Climate change may also mean that the adelgids will be more active when birds are flying south. Recently, the woolly adelgid has turned up in Ohio, Michigan, Vermont, New Hampshire, and Maine—approaching the northern limits of the hemlock range. Wherever it goes, it seems to get into every hemlock. It kills saplings before they can produce seeds, and so, in every place it arrives, it stops the hemlock species from reproducing. Many experts have concluded that the insect could kill nearly all the eastern hemlocks; if so, the species would essentially disappear from the wild.

ON DECEMBER 3, 2001, an arborist named Will Blozan discovered woolly adelgids on the branches of a wild hemlock in the Ellicott Rock Wilderness, on the Blue Ridge in South Carolina, near the extreme southern end of the hemlock range. No one had expected to see the insect this far south so soon. "It was a spear through the heart,"

Blozan told me. He phoned Rusty Rhea, an entomologist and the forest-health specialist for the Forest Service in Asheville, North Carolina. Rhea was surprised. He sent out a bulletin to all rangers in the area warning them to look for adelgids. Within two weeks, Rhea was getting reports. The insect had gone all over the mountains.

Will Blozan is a tall man in his thirties with dark blue eyes that can take on a guarded look, and he has a laconic way of speaking. Blozan has wide shoulders and powerful-looking hands, but his hands move with a sensitive, precise quality—they're the hands of a professional tree climber. He is the co-owner of a tree-care company called Appalachian Arborists, based in Asheville. (Arborists, who used to be known as tree surgeons, get around in trees using ropes.) He is also the president of the Eastern Native Tree Society, a small organization dedicated to discovering giant trees in the East. Since 1993, he had been spending his spare time exploring patches of old-growth forest in the Appalachians from New Hampshire to Georgia. Will Blozan became well known among tree biologists for having discovered and measured many of the tallest and largest trees in eastern North America. He often found them while he was bushwhacking through remote valleys in the southern Appalachians; he got into places that may not have had human visitors in years or decades.

In the Great Smokies in summer, the heat can be Amazonian. The land can slope sixty degrees, and in many places the undergrowth consists of a mesh of rhododendrons. "It's total suckage in there," Blozan said. " 'Rhodo wrestling' may be the appropriate term for movement in the Smokies."

When he found a big tree, he would get an estimate of its height using a laser device. Later, he would climb to the top, using ropes, and would send a measuring tape down along the trunk—this is the only way to determine the height of a tree to the nearest inch.

As he explored around, measuring tall trees, Blozan spent a lot of time around the southeastern tip of Great Smoky Mountains National Park, in a place called the Cataloochee Valley. It is a rain-forest wilderness, "well known for making you wet," Blozan said. The Cataloochee Valley is centered on a rumpled drainage dissected by hundreds of small upland coves and divided by ridges and mountains. A few hiking trails wander through the Cataloochee, but many parts of it are very difficult to enter.

Blozan eventually discovered that the Cataloochee Valley has the highest average tree height—more than 160 feet—of any watershed in eastern North America. The Cataloochee contains more than 80 percent of the world's tallest eastern hemlocks. It also contains the world's largest yellow poplar and four of the world's tallest white pines, including the tallest tree in eastern North America, a white pine that Blozan discovered in 1995 and named the Boogerman Pine. In January 2007, he found what turned out to be the world's tallest eastern hemlock, growing in a cove in the Cataloochee. He and Jason Childs, another arborist, climbed it and measured it with a tape, and got 173.1 feet. Blozan named the world's tallest hemlock Usis, which is the Cherokee word for "antler." "The Cataloochee is the epicenter of the eastern hemlock species," Blozan said. "The valley has the largest and best groves of eastern hemlock in the world." In effect, the Cataloochee Valley is the Notre Dame cathedral of the eastern forests.

By the summer of 2002, the woolly adelgid had been found in the Cataloochee. Few national parks have a forester on the staff, but the Great Smoky park does: Kristine Johnson. Kris Johnson is a slender woman in her fifties with a calm manner. Since 1990, she has been managing the park's efforts to beat back exotic invaders. "We currently have about a thousand sites in the park where exotic plants have gotten in, and we're dealing with ninety different species of invading organisms," Johnson told me—everything from Japanese stiltgrass to princess trees and fire ants. "We knew that sooner or later we would have the woolly adelgid. People around here have a saying: 'All the trouble comes from the North.' But we were still surprised by how quickly it got here."

The parasite may have been carried to North Carolina by people. In the summer of 2001, state nursery inspectors began finding infested hemlocks in nurseries in western North Carolina. The contaminated hemlocks had been imported into the state from areas where the bug was a problem. North Carolina inspectors ordered the nurseries to destroy their infested hemlocks. This was a money-losing deal for a business. "A speculation is that less-than-scrupulous nursery owners were unloading infested material on their customers," Kris Johnson said.

Now that the insect had arrived in the Great Smokies, what weapons were available to combat it? And how many hemlocks would need to be defended? It was clear that large numbers of hemlocks grew in Great Smoky Mountains National Park, but it wasn't immediately obvious, even to Kris Johnson, roughly how *many* there might be. She and her colleagues estimated, finally, that there might be 300,000 to 400,000 large hemlocks in the national park, not counting smaller, younger trees. Most of the hemlocks were tucked away in wilderness valleys, far from roads or trails.

Oil spray—the treatment that helps smaller trees in urban yards—wouldn't work in wilderness areas, where hundreds of thousands of large hemlocks would need to be drenched every year. There were two other promising options, though. Scientists at the University of Tennessee, funded in part by a private group, Friends of the Smokies, started a small lab for breeding a kind of lady beetle native to Japan that eats woolly adelgids. It was hoped that the beetles, released into the wild, would eat lots of adelgids, cutting down their numbers and eventually getting their population reduced to the point where hemlocks could survive the infestation.

Rusty Rhea, of the Forest Service, pushed the beetle strategy forward, and researchers began releasing the beetles. The lady beetles were tiny—the size of a sesame seed—and they initially cost about two and half dollars each; a cup of them cost thousands of dollars. When they were released at test sites, they had no measurable effect. But in 2007 (after years of test releases), a test in Banner Elk, North Carolina, in which different species of adelgid-eating beetles had been released over several years, had promising results. One type of beetle, from the Pacific Northwest, got established at the site and was eating adelgids, and the hemlocks there were looking better. Even so, there remained questions about whether and how quickly such results could be achieved on a large scale—if enough beetles could be bred and released and would multiply fast enough to save the hemlock forests that were dying or were under immediate threat. The beetles might work in the long run, but by then it might be too late for most hemlocks.

There was also an insecticide treatment, a chemical called imidacloprid, which is made by Bayer. Imidacloprid had to be mixed with water and injected into the soil around the root system of a hemlock. The chemical slowly moved into the foliage. When the adelgids sucked it into their bodies, they died. Imidacloprid is an artificial kind of nicotine. (Tobacco plants produce nicotine as a natural insecticide.) The injections were labor-intensive, but there was no good alternative: if imidacloprid was sprayed from the air, it would wipe out beneficial insects and wouldn't kill many adelgids—the insect's woolly coat sheds water.

The chemical had some advantages: it didn't migrate much through soil, so it would not be likely to spread widely into the environment, and it degraded quickly in sunlight. However, it was a toxic compound that could kill many grubs in the soil near the tree, as well as other insects feeding on the tree. It did not seem to affect vertebrates—frogs, salamanders, birds.

"I wouldn't want to see chemical treatment be the only way to save hemlocks, but nothing else is ready right now," Blozan said. "Either you get some invertebrate kill around the treatment site or you get an ecosystem collapse—that's the choice."

As soon as adelgids were found in the park, the Forest Service and Bayer began seeking Environmental Protection Agency permission to use the chemical in wild forests (it had already been approved for ornamental and landscape settings). The park treated ten old-growth hemlocks, as a test. Will Blozan's company, Appalachian Arborists, was hired to climb the trees and take samples of the foliage, to see how the chemical was moving through the tree. It can take a year or two for the benefits to become noticeable; some trees die anyway. "After treatment, the hemlock can look completely dead, but sometimes it will come back, and in three years it'll be vigorous," Blozan said. The ones that lived would need to be retreated every few years. The hemlocks would be like AIDS patients: they would never be free of the disease, though some might survive indefinitely on drugs. "What you get is a forest on life support," Blozan said. "But at least it can be kept alive while we hope for a cure."

Getting funding to fight the insect in the Great Smoky Mountains park was a byzantine process. The National Park Service, Kristine Johnson's employer, ran the park, but the Forest Service had responsibility for controlling pests in federal forests, including the national parks. The Forest Service is in the Department of Agriculture, while the National Park Service is in the Department of the Interior. Funding for insect control competes with other Forest Service needs, such as fighting forest fires. And the Forest Service appears to concentrate its pest-control efforts on trees that have commercial value—it had spent more than $100 million trying to get the emerald ash borer contained—and hemlocks aren't worth money.

Great Smoky Mountains National Park is the most-visited national park in the United States. More than nine million people pass through the park and experience its sights each year—that's more than twice the number of people who visit Grand Canyon National Park annually. Now the primeval rain-forest habitats of the Great Smoky national park were under grave threat. In 2003, Kristine Johnson asked for and got about $40,000 from the Forest Service to save the hemlocks in the Great Smoky Mountains park. In the next few years, the Forest Service spent about $15 million on research into ways to control the adelgid, but it spent very little to actually deploy the weapons that were available. By 2007, direct Forest Service funding for the park to fight the bugs was only $250,000 a year, with private donations increasing the total somewhat.

"The government is so damned slow," Blozan said. "Very little was done in the first two to three years."

Then, just as the insects appeared in the Great Smokies, Charles Taylor, a Republican North Carolina congressman who was the chairman of the appropriations subcommittee in charge of the national parks, began seeking $600 million from Congress to build a highway across Great Smoky Mountains National Park. The park was next to Congressman Taylor's district. Because the terrain in the Great Smokies was so rugged, the road would need to include three bridges, each likely to be longer than the Brooklyn Bridge—pork spanning a wilderness. Congressman Taylor argued that local residents would need to use the road and that it would bring jobs to his district. Environmentalists called it "the road to nowhere." The road went places in Congress, which appropriated $16 million to develop plans for it. This slug of funding tended to squeeze out other congressional appropriations for Great Smoky Mountains National Park. And none of the road money could be used for controlling the insects. In 1988, when Yellowstone National Park was devastated by forest fires, the federal government spent more than $100 million trying to put the fires out.

At any rate, the staff of Great Smoky Mountains National Park did what they could with the money they had. In 2003 and 2004, park employees treated hemlocks near public areas with imidacloprid—trees in campgrounds and along roads but not those deeper in the woods. The next year, the chemical was approved by the EPA for use in forests, and Kris Johnson and her colleagues designated special zones, called "hemlock conservation areas," where every hemlock would be treated. Will Blozan's company, Appalachian Arborists, won a contract and put a crew of five to work, while another crew, of eight, went to work under a park forester named Tom Remaley. The conservation areas totaled two square miles; Great Smoky park covers eight hundred square miles.

The biggest problem was carrying the water needed to mix with the chemical. The crews collected water from creeks in jugs, put the jugs in backpacks, and rhodo-wrestled their way up the mountainsides. A crew could treat between a hundred and four hundred hemlocks a day. At that pace, saving all the hemlocks in the national park was simply not possible. (Bayer later came up with a sort of a pill containing imidacloprid that could be tucked among the roots of a hemlock. The pill doesn't require water. As this is being written, the pill is being tested. If it works, crews carrying backpacks full of pills might be able to treat thousands of hemlock trees a day.)

In no other park were officials making the kind of effort that Great Smoky officials were. "Many other parks are 'monitoring the decline,' as I would put it, while they're implementing control in high-public-use areas," Johnson said. "I could put a hundred people to work treating hemlocks."

The woolly adelgid had not yet arrived in Cook Forest State Park, in northwestern Pennsylvania, which contained some of the richest old-growth eastern hemlock forest. "These parks should have a plan ready, and at the first sign of adelgids they should execute their plan," James Åkerson, the Shenandoah ecologist, said. A few million dollars—and a pill that works—would probably save the remaining fragments of old-growth hemlock forests. It wasn't clear that the government cared to spend the money, though.

While the parks were waiting for Washington, Appalachian Arborists was hired to treat hemlocks on private property using the soil-injection method. The Reverend Billy Graham had thousands of sick hemlocks at his religious training center near Black Mountain, North Carolina; Will Blozan saved them. "If you don't treat the tree, it will die," Blozan said, "and then you'll have to spend two or three thousand dollars having it removed." He also began treating another species, the Carolina hemlock. A very rare tree, the Carolina hemlock occupies a narrow range, primarily in North and South Carolina, where it grows on dry, rocky outcrops, on the lips of gorges and clinging to cliffs. There were thought to be about 520 Carolina hemlocks in South Carolina. At last count, there were exactly twelve of them known to be in Georgia. The Carolina hemlock looks like something out of a Chinese painting. It's a gnarled, wind-blasted thing with a mushroom-shaped top and downsweeping limbs flowing into space. Some of the smaller ones can be half a thousand years old. "The Carolina hemlocks are almost-beyond-words beautiful," Blozan said.

The state of South Carolina hired Blozan and his partners to try to save the state's 520 Carolina hemlocks, and the men treated nearly every specimen, often while rappelling down a cliff. Most of the Carolina hemlocks across the border in North Carolina were on national forest land, however, and were not being treated.

One day in August, I drove into the Cataloochee Valley with Will Blozan to see what had happened. In the back of his jeep were backpacks full of ropes and tree-climbing gear. We followed a dirt road that switchbacked down into the valley. It was a lush place, lined with meadows at the bottom, rising into ridges and coves blanketed with forest. The forest was streaked with gray areas, as if smoke filled it. We parked in a meadow and put on our packs. It was a hot day, and clouds were piling over the mountains. Blozan looked around. "The truth is,

I despise hiking," he said. "I don't do it unless there's a tree to climb somewhere on the hike." He wrapped a green bandanna around his head. We followed a trail that led into the woods along a creek called Rough Fork, crossing bridges made of single logs.

Big hemlocks, hundreds of years old, appeared. Sunlight seemed to blister its way through them. They were between 50 and 80 percent defoliated, but the national park crews had treated them, and many seemed to be alive, for now. "That one's looking better," Blozan noted, squinting at a hemlock that seemed half dead to me. He said that he had a map of the Cataloochee in his mind, with individual trees in it. "I've been all over these mountains. Even if I haven't seen a tree in ten years, I still know exactly where to find it in a crowd of trees," he said. "I've often wondered what a proctologist who's passionate about his work thinks when he sees a crowd of people."

He cut away from the trail, and we began bushwhacking up a slope. Here the trees were small, and patches of grass grew among them: the slope had been a cow pasture eighty years earlier. The slope ended on the knee of a ridge covered with rhododendrons. We followed the ridge upward, climbing steadily higher.

The trees got very big, and the forest seemed to get a lot darker. We had passed the edge of the old pasture and entered something like virgin American forest, a stretch of woods that had apparently never been logged. There were massive hardwoods—big yellow poplars, hickory trees, ash trees, mixed with an occasional sick-looking or dead hemlock. The ground was covered with all sorts of plants and shrubs—very high biodiversity, and the plants were all natives. No invading plants here. In various places there were woodland violets, early yellow violets, partridgeberry, masses of doghobble, wild lettuce, a lily called twisted rosy stalk, doll's-eyes, and Dutchman's-pipe. There was a brownish plant called squawroot that lives on rotting vegetation; bears eat squawroot in early spring after they come out of hibernation, and it purges their digestive systems.

The invaders were tiny or invisible to the naked eye. We passed hard, blackened stalks of wood: the remains of flowering dogwoods that had died a decade earlier from the invading dogwood fungus.

Here and there stood rotting beech trunks and dead standing hulks called snags. The beech trees had gone almost extinct in this part of the Cataloochee within the previous five years. One small beech tree was still alive, its bark dotted with flecks of white fungus. This was the European beech fungus spread by a European insect. The beech tree was still alive, but it was doomed, and it was the last of its kind visible in that part of the forest. Here and there lay huge, moundlike, rotting cylinders of wood: the fallen trunks of American chestnuts, which had most likely died during the 1930s and '40s, killed by the chestnut blight fungus, which drifted through in the air. Seventy years after dying, they still hadn't rotted away. The forest was a palimpsest, telling stories of loss and change.

As we went along, I found myself rhodo-wrestling. The rhododendrons won, and Blozan moved ahead. He seemed to slip through them without much effort.

Higher up, we crossed the ridge and looked down into a cove. It was drained by a creek named Jim Branch. As we moved downslope into the cove, sunlight began to flood the area, and the air grew hot and ovenlike. Around and above us extended ghosts, hemlocks that had been treated too late and were dead or mostly beyond saving. The ground was covered with surging plants, coming up in the light, including masses of stinging nettles. I couldn't see any adelgids; the parasites had died with their host. The air was filled with clouds of gray branches, like giant floating dust bunnies.

We stopped under a tall hemlock that glowed with green, a survivor in the cove. "This may be the healthiest hemlock in the park," Blozan said. It was known as Jim Branch No. 10, and it was 150 feet tall. One of the ten experimental trees that the park had treated in 2003, it had been treated again in 2005.

Blozan pulled the end of a climbing rope out of his pack and tied it to a cord he'd left strung in the tree. He used the cord to pull the rope into the tree, and he anchored the rope, making it safe for climbing. When you climb a tree, you begin by climbing up a rope into the crown of the tree. This is because most trees don't have branches near the ground that can be climbed on. The hemlock's lowest branch was about sixty feet above the ground.

Hemlock skeletons. Old-growth eastern hemlocks in the Cataloochee Valley killed by the hemlock woolly adelgid.
Will Blozan

Blozan put on a helmet and a tree-climbing harness and began ascending along the rope, using rope ascenders—mechanical devices that grab a rope and enable a person to climb up the rope without slipping down it.

I put on my helmet and harness and waited, watching Blozan. He got into the branches and kept moving upward until I could barely see him. Then I started ascending the rope, using ascenders.

Sixty feet above the ground, with Blozan climbing above me, I stopped and stood on a branch. (I was still attached to the rope, so that I couldn't fall.) Then I opened a bag and took out a complicated rig of ropes called a motion lanyard. The motion lanyard, also called a double-ended lanyard or a spider lanyard, is the principal tool used by some climbers for ascending to the tops of extremely tall trees, including hemlocks and redwoods. Here I will call it the spider lanyard.

The spider lanyard works much like Spider-Man's silk. You dangle in the air from the spider lanyard, which is attached to branches over your head. While your weight is suspended on the lanyard, you move upward by flinging alternating ends of it over branches overhead, getting it attached to successively higher branches. With a certain technique, using certain sliding knots, you can raise your body upward through the air, suspended on the rope, without touching the tree, or you can lower yourself, or you can hang motionless in midair on the spider lanyard, with your feet and hands touching nothing. More often, though, you hang on it with your feet lightly braced against the tree, for balance. Your life depends on the spider lanyard. If it is incorrectly attached, it can fail or a branch can break, and you will fall to the ground. A skilled tree climber can move from point to point in a tree while suspended entirely on ropes, not touching the tree with any part of his or her body.

Blozan was climbing rapidly above me, moving from branch to branch.

The tree was filled with a spicy tang, the scent of green hemlock, and it was covered with living things. There were rare dark-brown lichens called cyanolichens, which fix nitrogen straight from the air. They fertilize the canopy of old forests. There were beard lichens, horn lichens, shield lichens, and one called ragbag, which looks like rags in a bag. There were small hummocks of aerial moss, spiderwebs, insects associated with hemlock habitat. There were mites, living in patches of moss and soil on the tree, many of which had probably never been classified by biologists. The hemlock forest consists in large part of an aerial region that remains a mystery, even as it is being swept into oblivion by Mrs. Dooley's bug.

We stopped and rested at 130 feet. Blozan was standing on a small limb. "When these trees die, the nearby streams turn brown," he said. "The water gets full of tannic acid. As long as I've been coming here, these streams were crystal clear. Now they look like they're coming out of a bog." Many insects and fish that live in hemlock streams, such as the stone fly and the brook trout, are also threatened by sunlight and heat pouring into stream environments that were once shady and cool.

Blozan loosened his rope and climbed straight to the top. I followed him, moving more slowly. We spent a while lolling in the top of the tree. A bird landed near my face. It looked at me, hopped toward me, and let out a string of territorial cries. It was a red-breasted nuthatch, a species that feeds in hemlocks. Birds don't always seem to recognize a human in the top of a tree. Whatever this bird thought I was, it didn't seem to like having me there. A ruby-throated humming-

Will Blozan near the top of Jim Branch No. 10, seemingly the last healthy hemlock in Great Smoky Mountains National Park.

Richard Preston

bird began hovering around us, seemingly attracted to my red shirt. It throbbed off into the distance.

From the top of Jim Branch No. 10, we could see that the forest canopy was a ruin. The crowns of the dead trees were still encrusted with living material—a hemlock rain-forest canopy without the hemlocks. It was a scaffold of lichens and other organisms. The trees that harbored them had died so recently and so suddenly that they were all carrying on, for the moment, as if nothing had happened.

WHEN IT BECAME APPARENT that the eastern hemlock might nearly cease to exist, Blozan and his partners founded the Tsuga Search Project, an effort to identify and measure the world's tallest and largest eastern hemlocks before they were gone. They spent more than

A climber (Jason Childs) measuring Usis, the world's tallest eastern hemlock, soon after its discovery in 2007. It was alive at the time; it died a few months later. The climber is sitting in a tree-climbing harness, suspended from a rig of ropes similar to a spider lanyard.

Will Blozan

$100,000 of their own money on it. Brian Hinshaw, one of the partners, told me, "We just want to try to understand what we once had in these hemlocks." In the Cataloochee Valley, Blozan walked into groves where he found what had been among the world's tallest hemlocks. They were already dead, but he and his partners climbed the skeletons and measured them anyway. "The data are for someone someday," he said. In October 2007, Blozan and one of his partners, Jason Childs, went into a cove in the Cataloochee to check on the health of the world's tallest hemlock, Usis. Blozan had treated Usis with the chemical, and they wanted to see how it was doing. It was dead.

Three flagship species of migrating birds make their nests in hemlocks: the Blackburnian warbler, the black-throated green warbler, and the solitary vireo. In spring, they arrive in the Cataloochee before leaves come out on hardwoods; the evergreen hemlocks offer them cover, food, and a place to nest. No one knows what will happen to

A doomed canopy. Living masses of lichens clinging to dead hemlock branches in the moribund rain-forest canopy of the Cataloochee Valley.

Will Blozan

them when they arrive in the spring. Many other birds feed in hemlocks or nest in them, including the Acadian flycatcher, the Louisiana waterthrush, the winter wren, and the red-breasted nuthatch. The flying squirrel lives in hemlocks, and it feeds on fungi around their roots; the flying squirrel seemed to have gone into a decline. When an old hemlock falls, a world passes away. As for the Cataloochee Valley, most of the eastern hemlocks there were dead.

The Blood Kiss

1. Disturbed Forest

MONTHS LATER, when the epidemiologists finally arrived, they traced the threads of the horror back to one man, Patient Zero, who became known only by his initials, G.M. The threads converged on one little spot in the world. It was a sinuous patch of forest called Mbwambala. Mbwambala is a fragment of disturbed woodland about three miles long and half a mile wide that wanders along a small stream about six miles southeast of the city of Kikwit, in the Democratic Republic of the Congo. G.M. was forty-two years old at the time of his death. He lived with his extended family in Kikwit, in a family compound consisting of huts made of wattle. Each day, G.M. commuted on a bicycle to his place of work in the little forest of Mbwambala.

Kikwit is situated on the banks of the Kwilu River, 250 miles east of Kinshasa, the capital of Congo. Kikwit stands on a rolling, grassy plateau that is dissected by streams the color of chocolate milk. The streams meander through valleys filled with gallery forest, narrow, snakelike stands of trees nestled along the streams. Most of the plateau had once been covered with tropical rain forest, but 90 percent of it had been cut down in recent decades.

Nobody had a clear idea of how many people lived in Kikwit. Some experts felt that the city had a population of around 150,000, while others had a feeling that the city might contain closer to half a million people. Kikwit had grown into a sprawling agglomeration of houses made of wattle and cinder block, with roofs made of grass or sheets of metal. The city was a transportation center of a sort. Overland trucks passed through the city, moving along the Trans-Africa Highway, a system of dirt roads that crosses Africa from the Atlantic Ocean to the Indian Ocean. This is the same system of roads along which the human immunodeficiency virus, HIV, seems to have moved across Africa after its trans-species jumps from unidentified animal hosts—most likely monkeys and chimpanzees—into the human species.

Congo had been ruled for many years by the dictator Mobutu Sese Seko, who'd maintained control using a corrupt, chaotic army. Many things that had once worked reasonably well in Congo now did not. At one time, it had been possible to drive an automobile from Kikwit to Kinshasa in just four hours over a paved road. Now the drive to the capital lasted anywhere from twelve hours to several days, when the road was passable. It had developed ruts up to ten feet deep.

Kikwit had no running water, no sewage system, no telephone network, no newspaper, and no radio station. At night, the city went dark; there was very little electricity. The city's main hospital, Kikwit General Hospital, had a diesel generator. It also had an X-ray machine, but the hospital didn't have running water or toilets. It didn't even have bedpans: a family member would provide a clay jar to collect the patient's body waste. Many people who lived in Kikwit commuted by foot or bicycle to fields of cassava and maize outside the city. The people practiced slash-and-burn agriculture, moving their crops from place to place. The abandoned fields grew up in Christmas bush, an invading shrub from Florida and Central America. People also hunted small animals and gathered wood in the remaining patches of forest.

Just after dawn one day around Christmas 1994, G.M. got on his bicycle and pedaled through the city and down to the Kwilu River, and he crossed it on a bridge. He turned right and followed the river upstream through a valley. It was the rainy season. The weather was hot and wet, and the road was muddy; he threaded his way around puddles. After five miles he parked his bicycle, leaning it against a hollow stump near the road. Then he walked away from the road and away from the river, going uphill toward the east, following a footpath that went through thickets of Christmas bush and through fields of maize and cassava. He crossed a low ridge. The path descended. In about a mile, he came to a chocolate-colored stream. It wound among low hills through the forest of Mbwambala.

Mbwambala was a type of natural habitat that biologists call an ecotone. An ecotone is a transitional zone where two different ecosystems touch and mix. One sort of ecotone, for example, is the dividing line between wild forest and cleared agricultural land. Another ecotone is the meeting of sea with land, along a seashore. Ecotones are often richer in biodiversity. Living things find it easier to survive along edges, margins, boundaries, where different communities come together and mix, and where opportunities for feeding abound. Birds often congregate along ecotones—birds flock and feed along the edges of woods and along shorelines.

Mbwambala was a little ecotone, a narrow stretch of forest in the cleared country, running for a bit more than two miles along the stream. It contained groves of African corkwood trees. They had sprung up rapidly after the old, tall, primeval trees of Mbwambala had mostly been cut down. He walked along the stream, making his way under nightshade bushes the size of rhododendrons. There were small pigeonwood trees growing by the stream. Here and there, a few old forest trees remained—pale Tola trees and heavy, valuable Bomanga trees. These trees were 150 feet tall and maybe a century or two old. They were the remnants of a tropical forest canopy. Their crowns waved and flickered in the sunlight. In their tops lived bats, birds, insects, mites of the canopy—creatures that probably rarely came to the ground, if ever.

G.M.'s main profession was that of charcoalmaker. For years he had been cutting down trees in Mbwambala with an ax. He felled the trees and hacked up their branches and limbs and trunks into pieces, and made charcoal from them. The way he made charcoal was to dig a large pit, five feet deep, and fill it with pieces of wood. He set the wood on fire, then covered the burning wood in the pit with a layer of earth. The wood turned slowly into charcoal under the layer of earth.

Most cooking in Africa is done with charcoal. G.M. sold his charcoal to city people. That day, G.M. had just finished a session of charcoalmaking. He had recently removed a load of charcoal from his pit, and he now wanted to refill the pit. He spent the morning chopping down trees, lopping off their limbs, and moving the pieces of wood closer to his fire pit, in order to refill it with fresh wood.

What lived in the crowns of the trees he was chopping down no one really knows. Roughly half of all the species on earth are thought to live only in forest canopies. No one can say how many species exist on earth. Many biologists believe that most species on earth have not yet been identified and named. Most of these unnamed, undiscovered species (or life-forms) are viruses and bacteria. There are also thought to be many, many arthropods that have never been discovered. An arthropod is a small animal with an exoskeleton and jointed appendages—for example, an insect, a spider, or a shrimp.

Half a year later, when the disease hunters finally explored Mbwambala, they found a small, narrow hole that G.M. had dug during one of the last days of his life. The hole went down two feet among the roots of a tree. They wondered if he had dug up something in the hole and had eaten it. It might have been a tuber or a burrowing rodent with a nest of babies, or perhaps he had found a snake or some edible grubs there. Caterpillars are also a favorite food in Congo, especially a particular caterpillar that has a hard, shiny black head and a soft body and can grow up to five inches long. People in Congo roast it over a charcoal fire. Perhaps G.M. found some unusual wild caterpillars in the leaves of a tree he had cut down and ate them. He kept a few snares and traps in Mbwambala, for catching small animals, which he brought home to his family to eat. Perhaps he'd visited his snares, perhaps not; no one knows. Perhaps an animal bit him while he was taking it out of one of his snares; no one knows. The animals that turned up in his snares were mostly wild rats, including the giant African rat, which can be the size of a small dog. Some local people claimed, afterward, that G.M. had stolen an animal from someone else's snare.

Later in the day, he headed farther up the creek, deeper into Mbwambala. There, he visited a couple of maize fields he was tending. He had carved these fields out of the forest. During the heat of the day, he took a nap under a small shelter in one of his fields. Perhaps he was bitten by a spider or insect while he slept in the shelter. He returned to the city at dark. He had traveled twenty miles that day.

He would never visit the place again. Over the next few days, he began to feel unwell. He stayed home at his family compound in Kikwit. He ran a high fever; his eyes turned bright red. He got the hiccups, and they simply wouldn't stop. His face assumed a masklike appearance. He began defecating blood into his bed. His family took him to the Kikwit General Hospital, in the center of the city, where he died on January 13, of what people in the city would later call *la diarrhée rouge,* the red diarrhea.

2. Maternity Ward

BY LATE JANUARY, three members of G.M.'s family had died of *la diarrhée rouge.* Ten more members of his extended family, who lived in Kikwit and in surrounding villages, also came down with it. Some of them got endless hiccups, and all of them died. They, in turn, infected more people, and they all died.

Then it got into the Kikwit Maternity Hospital. This was a small collection of buildings in the south-central part of the city where pregnant mothers went to have their babies delivered. When a pregnant woman came down with it, the first sign was brilliant red eyes. The eyelids would eventually ooze blood, and the blood would stand on the edges of the eyelid in beaded-up droplets. The urine turned red—the kidneys were hemorrhaging; then the kidneys failed, and the person stopped urinating. The infected women in the maternity ward developed a masklike facial expression, and they became disoriented. Some had seizures. The disease was attacking the central nervous system. Some of them abruptly went blind. The skin became covered with a rash, a sea of tiny bumps, like goose bumps. The patients suffered from disseminated intravascular coagulation (DIC), in which the blood formed tiny clots throughout the body. At the same time, some of the patients were having hemorrhages, including bloody noses; in many patients, the stomach became distended and they began vomiting blood.

The illness invariably caused pregnant mothers to abort the children they were carrying; the fetus or baby was always either born dead or died shortly after birth. None of the babies of ill mothers survived.

During the delivery, the women experienced profuse, body-draining hemorrhages from the birth canal, and they died of hypovolemic shock. This is the shock that occurs when much of the blood has been drained from the body.

The doctors and nursing staff who worked in Kikwit Maternity Hospital did the best they could, but the hospital suffered from a shortage of basic medical supplies, such as rubber gloves. The doctors thought that they were dealing with an outbreak of dysentery.

On April 10, a medical technician who had been working with dying mothers at Kikwit Maternity Hospital came down with severe stomach pains. I will refer to him as the Maternity Technician. He went across town to the Kikwit General Hospital to get himself examined and treated. A doctor there suspected that the Maternity Technician had typhoid fever with peritonitis—a bacterial infection of the abdomen that is fatal if it isn't treated immediately. The doctors at Kikwit General Hospital put the Maternity Technician into surgery.

A group of Italian nuns worked in the hospital as nursing sisters. They were known as the Little Sisters of the Poor, and they came from a convent in Bergamo, Italy.

One of the nuns, Sister Floralba Rondi, assisted two surgeons and a nursing team in the operating room on the day they operated on the Maternity Technician. The lead surgeon made a vertical cut down the center of the technician's abdomen, opened him up, and looked into his abdominal cavity. They were expecting to see pus. Pus occurs with a bacterial infection. There was no pus; there were no bacteria. (Viruses are not bacteria.) The surgical the team took out the man's appendix.

The next day, however, the Maternity Technician grew worse. His abdomen became very swollen and distended. Wanting to see what was causing the distension, the doctors inserted a hypodermic syringe into his abdomen and extracted a sample. The syringe filled up with blood. The blood had a runny, homogenized appearance. It wouldn't coagulate.

The doctors brought him back into the operating room and opened him up through the same incision as before, in an effort to find the source of his bleeding and stop it. They couldn't find any source for the bleeding. The blood seemed to be coming directly out of his organs, as if from a squeezed sponge. They sewed him back up. By the time the surgical team had finished the surgery, the team members were probably smeared with the Maternity Technician's blood and probably had it all over their hands. Some or all of the team members performed the surgery without gloves, with bare hands.

Two days later, the Maternity Technician died.

In the next ten days or so, nearly every member of the team that had operated on the Maternity Technician also died, including Sister Floralba and two surgeons. Other medical staff who had been caring for the Maternity Technician, including Sister Dinarosa, died as well. At this point, it was clear that there was a dangerous disease loose in the hospital. The doctors wanted to get the word out and get help. There was no telephone at the hospital, but the hospital had a communication radio. Every evening at the same time, the surviving nuns sent out a radio bulletin, reporting on the events of the day, the deaths that had occurred. This message was relayed by fax every day to the convent of the Little Sisters of the Poor, in Bergamo, Italy. The nuns in Bergamo were becoming increasingly alarmed about the deaths of their sisters in Kikwit.

As the news got out that the disease killed practically everyone who got it, the city of Kikwit went into a panic. Almost all of the patients who were in Kikwit General Hospital fled back to their homes, fearing the disease in the hospital. Some of them went to the villages surrounding Kikwit. Because of the bleeding and the high rate of mortality, the doctors began to believe that they were dealing with Ebola.

They set up an isolation ward in Pavilion Three of the Kikwit General Hospital. They brought in thirty patients who were suffering from the red diarrhea and placed them in the beds. The mattresses soon became soaked with blood and filth, and the floors became slippery with blood.

One of the physicians, Dr. Mpia A. Bwaka, volunteered to stay in Pavilion Three with the patients. By this time, it was fairly obvious to Dr. Bwaka that this decision meant that he would probably die. He was helped by three male nurses who also volunteered to stay with the patients, even though they knew that they would probably get the disease and die. The nurses' names have not been recorded in the medical literature; they were local men from Kikwit who were earning next to nothing for their work. Dr. Bwaka wasn't getting paid, either. The hospital's staff had not received their salaries in several months; economic conditions in Congo were very bad.

Dr. Bwaka and the three men gathered up the hospital's small supply of rubber gloves and took them into Pavilion Three. They wore the gloves sparingly, washed them, and reused them. They wore cloth surgical gowns and handmade masks woven locally from cotton. They slept in the pavilion with their patients. As the patients died, the corpses were left in the beds or were placed on the floor, to make room for more people being brought into Pavilion Three. Dr. Bwaka and his team were working literally up to their elbows in blood, black vomit, and excrement the color of beet soup. The team didn't even have running water to rinse the floors of Pavilion Three. By now, hundreds of people in Kikwit and the surrounding towns were dying of it. It had all come from one man who had gone into the forest in Mbwambala and come into contact with some wild creature there.

3. Identification

IN PAVILION THREE of the Kikwit General Hospital, Dr. Mpia Bwaka collected samples of blood from fourteen of his patients. Somebody drove the samples over the terrible road to Kinshasa. From there, the blood samples were flown to a laboratory in Belgium. The Belgian scientists, fearing that the blood might be dangerous, sent them along to the Centers for Disease Control and Prevention in Atlanta—the CDC. The fourteen tubes of blood from Pavilion Three ended up in the Biosafety Level 4 hot zone of the Special Pathogens Branch, where a researcher named Ali S. Khan, working along with several colleagues, identified Ebola virus in all fourteen of the test tubes. It was a new type of Ebola virus, and it would eventually be named Ebola Kikwit.

As it happened, just at that time, *another* new type of Ebola was causing an outbreak in Ivory Coast, in West Africa. This type of the virus would eventually be named Ebola Ivory Coast. Thus two outbreaks of different types of Ebola occurred almost simultaneously in different places, which deepened the mystery over the origin of Ebola. Why and how was Ebola emerging in different places? A medical team from the World Health Organization, in Geneva, was preparing to fly to Ivory Coast, in West Africa, to investigate.

The WHO Ebola Ivory Coast team was led by Bernard Le Guenno, a scientist from the Institut Pasteur in Paris, and by Pierre Rollin, a French virologist who was then stationed at the CDC in Atlanta. But with the large outbreak happening in Congo, which needed immediate attention, Bernard Le Guenno and Pierre Rollin were sent to Congo instead, where they joined a ten-member WHO team of doctors from France, Congo, the United States, and South Africa.

The Ivory Coast Ebola case had occurred about a month before G.M. got sick—he was the first person known to have Ebola Kikwit. In Ivory Coast, a woman scientist from Switzerland (whose name has never been publicly disclosed) was studying a troop of wild chimpanzees in Taï National Park. The Taï Forest was one of the last pristine rain forests in West Africa. The troop of chimpanzees became infected with a mysterious disease, and many of them died. The Swiss woman, extremely concerned about her chimps, dissected one of the dead animals, trying to find out what had killed it. Soon afterward, she developed a rash and became severely ill, and she began having hemorrhages. She developed the symptoms of Ebola virus.

For unknown reasons, Ebola had been getting into chimpanzees. Chimps and other great apes, such as bonobos and gorillas, are probably not natural hosts of Ebola virus. This is because Ebola makes the apes extremely sick—as sick as humans become with the virus. (The western gorilla is presently very threatened by Ebola virus, and many gorillas have died in Congo from outbreaks of Ebola among them. No one knows how the gorillas are getting Ebola, but some wildlife biologists fear that Ebola could help cause the western gorilla's extinction.) The fact that Ebola is exceedingly lethal in monkeys and apes means that the natural host of Ebola is probably not a monkey or ape—those animals haven't developed resistance to it. But somehow, the chimpanzees of the Taï Forest were coming into contact with Ebola's host.

The Swiss woman was flown on a commercial airliner to Switzerland for treatment. Her doctors in Switzerland did not realize that she was infected with Ebola. They suspected that the illness was dengue hemorrhagic fever, a virus carried by mosquitoes. Nevertheless, she survived, and no one else in the hospital in Switzerland got sick.

The Taï chimps ate all sorts of things. They hunted colobus monkeys and ate them raw, tearing them apart, a bloody process. Possibly the chimps were catching Ebola from dead monkeys; the monkeys might have been catching Ebola from some *other* creature they were hunting.

4. Blood Kiss

WHILE EBOLA was breaking out in Kikwit, I spoke with a doctor named William T. Close, who had lived in Congo (then Zaire) for sixteen years. When he was in Zaire, Bill Close rebuilt and ran the Mama Yemo General Hospital, a two-thousand-bed facility in the capital. When Ebola broke out for the first time, in 1976, Close went to Zaire and helped coordinate the medical effort to deal with the virus, and advised the Zairian government. Years later, during the Ebola Kikwit outbreak, he acted as a liaison between Congolese government officials and doctors from the CDC in Atlanta.

"In 1976, when Ebola broke out in Yambuku that first time, there was a nun, Sister Beata, who died of Ebola," Close recalled. "There was a priest, Father Germain Lootens, who gave her the last rites as she died. She had a terrible fever, sweat was pouring down her face, and bloodstained tears were running down her face. Father Lootens took out his handkerchief and wiped the sweat from her forehead and the bloody tears from her face. Then, unthinkingly, he took the tearstained handkerchief and wiped the tears from his own face with it—he had been crying, too. A week later, he came down with Ebola, and a week after that he was dead."

Now, Close had been hearing reports that some members of the medical staff of Kikwit General Hospital—Dr. Bwaka and his nurses—had continued to care for Ebola patients despite the grave risks to themselves. "Those hospital staff people have gone into that hospital to work knowing that they may die," Close said. And the doctors and nurses in Kikwit were working without basic medical supplies. "The greatest need in Kikwit right now is for rubber aprons to protect the doctors and nurses, because the blood and vomit is soaking through their operating gowns," he said. "This is a huge, lethal African hemorrhagic virus. We all sort of feel that Ebola comes out of its hiding place when something occasionally alters the very delicate balance of the ecosystems, in a tropical region where things grow as they would in a petri dish. But if there are lessons to be learned here, they are human lessons. This is about people doing their duty. It's about doctors doing what has to be done, right now, without a whole lot of heroics. Have you ever been petrified with fear? Real fear? Possessed by naked fear, where you have no hope of control over your fate? If you're a medical worker, when the die is cast, the fear goes away, and you do what you have to do—you get to work. That's what's happening with the medical people in that hospital right now. There are things happening in Kikwit . . ." He paused. "Magnificent human things. . . . How can I explain this? There was another incident in 1976, also in Yambuku. One of the doctors—he was a Belgian named Jean-François Ruppol—delivered a baby in the middle of it all." Ebola has a profound effect on pregnant women: they hemorrhage profusely and abort the fetus, which itself is infected with Ebola. "There were people dying of Ebola all around in that room in the hospital, and there was a woman in childbirth. She was Dr. Ruppol's patient, and her baby was his patient, too. The baby was stuck—too big for the birth canal." The woman had a high fever, she was terribly ill, but her baby had to be delivered, even if it was infected with Ebola. "So he performed the Zarate procedure on her," Close said.

"What's that?" I asked.

"The Zarate procedure? It's a simple and rather crude but very effective way of enlarging the outlet to remove the baby. With a knife, you split the pubic symphysis."

"The what?"

"The front of the pelvis. The pelvic bones," he said. It's a hard, bony spot, and you can feel it, just above the pubic area, he said. "You split the bones there. You press a scalpel through cartilage. The bones go *pop* and the pelvis springs open, and you pull the baby out. The hospital had run out of anesthetics. So he did the Zarate procedure on the woman without giving her an anesthetic."

"My God."

"She was conscious. By the time he got the baby out, the baby had stopped breathing. The baby was in breathing arrest and drenched with the woman's blood. He put the baby's mouth to his mouth and gave the baby mouth-to-mouth resuscitation. The baby started to breathe. He pulled away, and his mouth and face were smeared with blood. There was a nurse standing by. When she saw his face she said, 'Doctor, *do you realize what you've done?*'

" 'I do now,' he said."

5. Seeking the Ghost

When the WHO team arrived in Kikwit, they found Dr. Mpia Bwaka working alone in Pavilion Three with only two nurses—the third nurse had died of Ebola a few days earlier. Dr. Bwaka seemed to be all right. The WHO team had brought medical supplies, including jugs of bleach. They washed the ward with the bleach, rinsing the blood and feces off the floor. The team members put on double rubber gloves, waterproof gowns, masks, and goggles, and distributed the same equipment to Dr. Bwaka and his nursing staff. They wrapped the mattresses (which were blood-soaked) in plastic covers. Afterward, Ebola patients were placed directly on the plastic, without sheets. A Belgian team from Doctors Without Borders arrived a few days later, and put up white Tyvek sheets around Pavilion Three, as a sort of crude barrier to keep the virus inside the pavilion; the Belgian team also brought water for the hospital. Dr. Bwaka continued to work in the Ebola ward. It was so hot that the goggles fogged up, so the medical workers often didn't wear them. One day, a nurse forgot himself momentarily and wiped his eyelid with his gloved fingertip, which was contaminated with Ebola blood. He died of Ebola.

But by the time the teams arrived in Kikwit, the outbreak was fad-

Sister Beata waving good-bye on the Ebola River. In 1976 she would die of Ebola while a priest, Father Lootens, wiped bloody tears from her face with his handkerchief and then unthinkingly wiped his own face with the bloody handkerchief, a mistake that sealed his doom.

Courtesy of William T. Close

ing away. What really ended it was the fact that the virus had killed a third of the doctors in the city. Once the medical system collapsed, people didn't go to the hospitals where the virus had spread. The outbreak burned itself out. Dr. Mpia Bwaka survived.

In the following months, a team of epidemiologists and zoologists led by Herwig Leirs, an ecologist at the Danish Pest Infestation Laboratory in Lyngby, Denmark, fanned out into the countryside around Kikwit and began trapping animals and birds and testing their blood. They were trying to find a species of animal that was either infected with Ebola or had antibodies to Ebola in its bloodstream, which would suggest that the animal was a natural carrier of the virus. They set out traplines and mist nets all through the forest of Mbwambala, and in other places in the countryside around the city. In the end, they collected slightly more than three thousand specimens. Most of them were mammals. About ten percent of the specimens were birds, and a few of them were reptiles and amphibians. Most of the mammals were rodents, and there were a number of bats. But they also collected wild African cats, as well as wild red pigs, pangolins, and elephant shrews. Not one of the specimens turned up positive for Ebola virus. Not one.

The Danish team didn't look at any insects. Insect biodiversity in tropical Africa is enormous and unfathomed—many species of insects in Congo have never been identified or given names. A collecting team led by Paul Reiter of the CDC went around Kikwit and the countryside and collected thirty-five thousand arthopods—insects, ticks, sand flies, fleas, lice. They collected many bedbugs from around the city. For some reason, they didn't catch any spiders or scorpions. They also didn't report collecting any mites. (Mites are very small arthropods that are very difficult to see and collect.) Mites can live in hair follicles or on the skin of an animal or person, as well as in soil. The CDC arthropod team didn't find any trace of Ebola in any of the thirty-five thousand specimens. No Ebola in a single bug.

It left the mystery unsolved. In what creature does Ebola make its everyday home? One interesting question about Ebola is this: Why aren't humans infected more often with Ebola? Why are the outbreaks actually quite rare? If Ebola lives in some common animal or insect, then people should become infected more frequently. Possibly Ebola lives in primeval rain-forest canopies, in some creature that exists high above the ground in the remains of an ancient forest ecosystem. When a forest is disturbed—when trees are chopped down—people come in contact with the canopy and all that lives there. Perhaps the first man with Ebola in Kikwit, G.M., cut a tree down, then touched or ate a bat, bird, or insect that lived only in the tops of trees. Or perhaps he got Ebola from something that had lived underground, something he found in the small hole he dug that day in Mbwambala. He was dead, and many members of his family—who might otherwise have been able to recall details of his activities during the days when he became infected—were dead, too. Ebola kills the witnesses to its appearance. There were hints that some type of bat might be the natural host of Ebola. In laboratory tests, Ebola virus has been able to infect certain kinds of bats without making them sick. The bat's immune system is resistant to Ebola, which suggests bats may be carriers of the virus. Even so, no wild bats have ever been found with Ebola in them.

Bats have very unusual wingless parasitic flies that live on them, sucking their blood. These bloodsucking bat flies, called strebelid flies, crawl from bat to bat while the bats are hanging in roosts. The flies might transmit Ebola among the bats. Does Ebola live in wingless flies crawling on bats? Nobody knows.

This is a story with no end. Recently, I called Dr. William Close, to see how he was doing. He lives in Big Piney, Wyoming.

"That Belgian doctor," I said. "The one who got the Ebola-infected blood all over his face? How long did he survive?"

Close began chuckling. "More than thirty years, so far. I just talked with him yesterday. Jean-François Ruppol. He's a great friend of mine. He lives in Belgium now."

I could hardly believe it. How could anyone survive an Ebola exposure like that?

Not long afterward, I received a series of pleasant e-mails from Dr. Jean-François Ruppol. He had written down, in French, some of his recollections of the first Ebola outbreak, in Yambuku, near the Ebola River, Congo, in 1976. Ruppol went to Yambuku three times during the outbreak, seeking to understand the virus and get it under control. (At the time, Ebola virus did not yet have a name.) Here, in Ruppol's words, is what happened:

> *Ma première nuit à Yambuku fut calme. . . .* My first night in Yambuku was calm, but around five o'clock in the morning, a nursing sister woke me, banging on the door of the room I was occupying. A woman had just been brought in who had been in labor for a full day, and her situation didn't look good. I have to admit that I was a little nervous. For one thing, I didn't want to go into the hospital or the maternity ward, where there had been numerous sick patients and where the virus might still be present in patches of blood and soiled sponges that were scattered all about. For another thing, practically all of the male and female nurses had died, and the sur-

vivors had fled. Was the woman they had just brought in contaminated?

At this point, I asked a nun if they could put a kitchen table on the building's porch. We put the pregnant woman on the table, after we had donned protective gear (gown, cap, mask, gloves, etc.). I wanted to take all the necessary precautions, the same ones I had ordered others to use during this epidemic.

In the course of my examination, I came to the following conclusions:

- The woman was at the end of her rope.
- The fetus was presenting in a dangerous way. If I remember correctly, the fetus was stuck sideways, making birth impossible.
- The fetus was in extremis.

We had to act quickly. But a caesarean was impossible because of the dangers in the operating room, the blood and foul sponges, and because of the absence of qualified personnel. Therefore I decided to utilize a technique that I had occasionally practiced in Kimpangu, the symphysisiotomy [the Zarate procedure]. It consisted of cutting the cartilage at the pubic symphysis, and then spreading the legs to open the pelvis and favor the passage of the fetus.

Getting the help of two people to hold the mother's knees and legs, I performed the Zarate procedure under a local anesthetic, and I reached in and turned the fetus around inside her, in order to deliver it bottom-first.

Illuminated by flashlights and an electric light from a generator, the maneuver went well, but once the baby was delivered and the umbilical cord cut, the baby would not breathe despite various attempts to wake it up. Then, pushed by habit (or instinct, perhaps?), I took down my mask and practiced gentle mouth-to-mouth resuscitation. At that very moment I got a terrible shock: I realized that if the woman was infected with the virus, then I had just condemned myself to death. This was because we knew the virus was transmitted in all the

> secretions and fluids of the body. Even so, the child was revived and the mother seemed to be doing all right. It's hardly necessary to add that I spent the next forty-eight hours keeping a very close watch on the health of the mother and baby. Oof! They weren't contaminated, and I was alive. This was the only time in my medical career when I was not just afraid, but felt and lived real terror. . . .

Ruppol had lost his sense of self-protection during the emergency, but had gotten lucky. The mother hadn't had Ebola.

Close thought it was just typical of the way doctors can forget themselves when a patient is in trouble. It didn't give him any confidence, though, that the doctors had the situation with emerging viruses and microbes under control. "In the battle between the doctors and the bugs," he remarked, "in the long run, I'd put my money on the bugs."

The Human Kabbalah

"Craig Venter is an ass. He's an idiot. He is a thorn in people's sides and an egomaniac," a senior scientist in the Human Genome Project said to me one day. The Human Genome Project was an ongoing nonprofit international research consortium that had been working to decipher the complete sequence of nucleotides, or letters, in human DNA. The human genome is the total amount of DNA that is spooled into a set of twenty-three chromosomes in the nucleus of every typical human cell. (There are two sets of chromosomes, for a total of forty-six chromosomes in each human cell.) This entire package of DNA in every cell is sometimes referred to as the book of human life. Most scientists agreed that deciphering it would be one of the great achievements of our time. The stakes, in money and glory, to say nothing of the future of medicine, were huge and incalculable.

In the United States, most of the money to pay for the Human Genome Project had been coming from the National Institutes of Health, the NIH. The project was often referred to, in a kind of shorthand, as the "public project," to distinguish it from for-profit enterprises like the Celera Genomics Group, of which Craig Venter was the president and chief scientific officer. "In my perception," said the scientist who was giving me the dour view of Venter, "Craig has a personal vendetta against the National Institutes of Health. I look at Craig as being an extremely shallow person who is only interested in Craig Venter and in making money. Only God knows what those people at Celera are doing."

What Venter and his colleagues were doing was preparing to announce that they had placed in the proper order something like 95 percent of the readable letters in the human genetic code. They were referring to this milestone as the first assembly. They had already started selling information about the human genome to subscribers. The Human Genome Project, largely in response to Craig Venter and the corporate effort to read the human book of life, was also on the verge of announcing a milestone. Its scientists were calling their milestone a "working draft" of the genome. They were claiming it was more than 90 percent complete, and they were making the information available to anyone, free of charge, on a database called GenBank. Both images of the human genome—Celera's and the public project's—were becoming clearer and clearer. The book of life and death was opening, and we held it in our hands.

A HUMAN DNA MOLECULE is about a meter long. It is about a twenty-millionth of a meter wide—the width of twenty hydrogen atoms. It is shaped like a twisted ladder. Each rung of the ladder is made of one of four nucleotides—adenine, thymine, cytosine, and guanine. The DNA code is expressed in combinations of the letters A, T, C, and G, the first letters of the names of the nucleotides. The human genome contains at least 3.2 billion letters of genetic code. This is about the number of letters in three thousand copies of *Moby-Dick*.

Perhaps three percent of the human code consists of genes. Genes hold the recipes for making proteins. Human genes are stretches of between a thousand and fifteen hundred letters of code, often broken into pieces and separated by long passages of DNA that don't code for proteins. It is believed that there are about twenty-five thousand genes in the human genome. Much of the rest of the genome consists of blocks of seemingly meaningless letters, gobbledygook. These sections are referred to as junk DNA, although it may be that we just don't understand the function of the apparent junk.

The conventional route for announcing scientific breakthroughs is publication in a scientific journal. Both Celera and the Human

Genome Project were planning to publish annotated versions of the human genome as soon as possible. Although the two sides looked like armies maneuvering for advantage, the leaders of the Human Genome Project had always denied that they were involved in some kind of competition with Craig Venter.

"They're trying to say it's not a race, right?" Venter said to me, in a shrugging sort of way. "But if two sailboats are sailing near each other, then by definition it's a race. If one boat wins, then the winner says, 'We smoked them,' and the loser says, 'Eh, we weren't racing—we were just cruising.' "

I first met Craig Venter on a windy day in the summer of 1999, at Celera's headquarters in Rockville, Maryland, a half-hour drive northwest of Washington, D.C. The company's offices and laboratories occupied a pair of five-story white buildings with mirrored windows, surrounded by beautiful groves of red oaks and yellow poplar trees. One of the buildings contained rooms packed with row after row of DNA-sequencing machines of a type known as the ABI Prism. The other building held what was said to be the most powerful civilian computer array in the world. The Celera supercomputer complex was of considerable interest to Gregory and David Chudnovsky, the mathematicians who had used a homemade supercomputer to calculate the number pi, and who ended up meeting with Craig Venter and his staff, talking with them about the design of supercomputers and software used in sequencing the human genome. Venter's supercomputer complex was surpassed only, perhaps, by that of the Los Alamos National Laboratory, which is used for simulating nuclear bomb explosions.

The computer building at Celera also contained the Command Center. This was a room stuffed with control consoles and computer screens. The Command Center was manned all the time. It monitored the flow of DNA inside Celera. The DNA was flowing through the machines twenty-four hours a day, seven days a week.

That hot summer day, Craig Venter moved restlessly around his office. There had been a spate of newspaper stories about the race to decode the complete genome, and about the pressure Celera was putting on its competitors. "We're scaring the crap out of everybody, including ourselves," he said to me.

Venter was fifty-three at the time. He had an active, cherubic face on which a smile often flickered. He was bald, with a fuzz of short hair at the temples, and his head was usually sunburned. He had bright blue eyes and a soft voice. That day, he was wearing khaki slacks and a blue shirt, New Balance running shoes, a preppy tie with small turtles on it, and a Rolex watch. Venter's office looked out into stands of trees; leaves were spinning on branches outside the windows, flashing their white undersides and promising thunderstorms. Beyond the trees, a chronic traffic jam was occurring on the Rockville Pike. Celera was in an area along a stretch of Interstate 270 known as the biotechnology corridor, which was dense with companies specializing in the life sciences, and billions of dollars in venture capital were embedded in bioenterprises all around Celera.

Celera Genomics was a part of the PE Corporation, which had been called Perkin-Elmer before the company's chief executive, Tony L. White, split the business into two parts: PE Biosystems, now called Applera, which made the DNA sequencing machine called the Prism, and Celera. Venter owned five percent of Celera's stock. It had been trading, often violently, on the New York Stock Exchange. The stock had been tossed by waves of panic selling and panic buying. That particular summer day, the stock market was valuing Celera at around three billion dollars. Craig Venter's own net worth had been slopping around by five or ten million dollars a day in either direction, like water going back and forth in a bathtub.

"Our fundamental business model is like Bloomberg's," Venter said. "We're selling information about the vast universe of molecular medicine." Venter hoped, for example, that one day Celera would help analyze the genomes of millions of people as a regular part of its business. This would be done over the Internet, he felt—and, having decoded individual patients' DNA, the company would then help design or select drugs tailored to patients' particular needs. In recent times, genomics has been moving so fast that it's possible to think that pretty soon you will be able to walk into a doctor's office and have your own genome read and interpreted. It could be stored in a smart card. (You would want to keep the card in your wallet, in case you landed in an emergency room. But you wouldn't want to lose it, because if thieves got your DNA sequence, they might *really* be able to clone you.) Your doctor would read the smart card, and it would show your total biological-software code. Your doctor would be able to see the bugs in your code. The bugs are genes that make you vulnerable to certain diseases; everyone has bugs in their code. If you became sick, doctors could watch the activity of your genes, using so-called gene chips, which are small pieces of glass containing detectors for every gene. Doctors could track how your body responded to treatment. All your genes could be observed, operating in an immense symphony.

Venter stopped moving briefly, sat down in front of a screen, and tapped a keyboard. A Yahoo! quote came up. "Hey, we're over twenty today," he said. Meanwhile, I was standing in front of a large model of Venter's yacht, the *Sorcerer,* in which he'd won the 1997 Transatlantic Challenge in an upset victory—it was the only major ocean race he'd ever entered. "I got the boat for a bargain from the guy who founded Lands' End," Venter said. "I like to buy cast-off things on the cheap from ultrarich people."

Venter went into the hallway, and I followed him. Celera was renovating its space, and tiles were hanging from the ceiling. Some had fallen to the floor. Black stains dripped out of air-conditioning vents, and sheets of plywood were lying around. Workmen were Sheetrocking walls, ripping up carpet, and installing light fixtures, and the smell of paint and spackle drifted in the air. We took the stairs to the basement and entered a room that held about fifty Prism DNA-sequencing machines. Each Prism was the size of a small refrigerator and had cost three hundred thousand dollars. Prisms were the fastest DNA sequencers on earth. At the moment, they were reading the DNA of the fruit fly. This was a pilot project for the human genome. The machines contained lasers, which were used for reading the letters in DNA. Heat from the lasers seemed to ripple from the machines. The lasers were shining light on tiny tubes through which strands of fruit-fly DNA were moving, and the light was passing through the DNA, and sensors were reading the letters of the code. Each machine had a computer screen on which blocks of numbers and letters were scrolling past. It was fly code.

"You're looking at the third-largest DNA-sequencing facility in the world," Venter said. "We also have the second largest and the largest."

We got into an elevator. The walls of the elevator were dented and bashed. Venter led me into a vast, low-ceilinged room that looked out into the trees. This was the largest DNA-decoding factory on earth. The room contained 150 Prisms—forty-five million dollars' worth—and more Prisms were due to be installed any day. Just below the ceiling, air ducts dangled on straps, and one wall consisted of gypsum board.

Venter moved restlessly through the unfinished space. "This is the most futuristic manufacturing plant on the planet right now," he said. Outdoors, the rain came, splattering on the windows, and the poplar leaves shivered. We stopped and looked over a sea of machines. "You're seeing Henry Ford's first assembly plant," he said. "What don't you see? People, right? There are three people working in this room. A year ago, this work would have taken one thousand to two

J. Craig Venter recently, in a sailboat.

Evan Hurd/Getty Images

thousand scientists. With this technology, we are literally coming out of the dark ages of biology. As a civilization, we know far less than one percent of what will be known about biology, human physiology, and medicine. My view of biology is 'We don't know squat.' "

Some observers thought the company could fail. It was burning through at least $150 million a year. This flow of money going out of Celera was what venture capitalists called the "burn rate" of a start-up company—its negative cash flow, its consumption of money without (yet) producing a cash return on the investment. Who, I wondered, would want to buy the information the company was generating, and how much would they pay for it? "There will be an incredible demand for genomic information," Venter assured me. "When the first electric-power companies strung up wires on power poles, there were a lot of skeptics. They said, 'Who's going to buy all that electricity?' We already have more than a hundred million dollars in committed subscription revenues over five years from companies that are buying genomic information from us—Amgen, Novartis, Pharmacia & Upjohn, and others. After we finish the human genome, we could do the mouse, rice, rat, dog, cow, corn, maybe apple trees, maybe clover. We could do the chimpanzee."

ONE DAY AT CELERA'S HEADQUARTERS, I was talking with a molecular biologist named Hamilton O. Smith. Smith, an extremely distinguished figure in the history of molecular biology, won a Nobel Prize in 1978 as a codiscoverer of restriction enzymes, which are used to cut DNA in specific places. Scientists use these enzymes like scissors, chopping up pieces of DNA so that they can be studied or recombined with other pieces of DNA. Without the DNA-cutting scissors that Hamilton Smith discovered, there would be no such thing as genetic engineering or molecular biology. Most people in his field who knew him called him Ham Smith.

Ham Smith was in his late sixties. He stood six feet five inches tall. He had a shock of stiff white hair and a modest manner. Working for Celera, he seemed to be a putterer, knocking around in a sophisticated lab while helping the company decode the human genome.

"Have you ever seen human DNA?" Ham Smith asked me, as he poked around his lab.

"No."

"It's beautiful stuff."

He brought me over to a small box that sat on a countertop. It held four small plastic tubes, each the size of a pencil stub. "These four tubes hold enough human DNA to do the entire human genome project," Smith said. "There's a couple of drops of liquid in each tube."

He lifted up one of the tubes and turned it over in the light to show me what DNA looks like to the naked eye. A droplet of clear liquid moved back and forth in the tube. It was the size of a dewdrop. Then he held up a glass vial and rocked it back and forth; a crystal-clear, syrupy liquid oozed around in it. "That's long, unbroken DNA," Hamilton Smith said. He'd extracted it from human blood—from white blood cells. "This liquid looks glassy and clear, but it's snotty," he went on. "It's like sugar syrup. It really is a sugar syrup, because there are sugars in the backbone of the DNA molecule. Watch this."

Smith picked up a pipette, a handheld device with a hollow plastic needle in it, used for moving tiny quantities of liquid from one place to another. His hands were large, but they moved with precision. Holding the pipette, he sucked up a droplet of DNA mixed with a type of purified salt water called buffer. He held the drop in the pipette for a moment, then let it go. The droplet drooled. It reminded me of a spider dropping down a silk thread.

"There the DNA goes, it's stringing," he said. "The pure stuff is gorgeous."

The molecules were sliding along one another, like cooked spaghetti falling out of a pot, causing the water to string out. "It's absolutely glassy clear, without color," he went on. "Sometimes it pulls back into the tube and won't come out. I guess that's like snot, too, and then you have to almost cut it with scissors. The molecule is actually quite stiff. It's stiff like a plumber's snake. It bends, but only so much, and then it breaks. It's brittle. You can break it just by stirring it."

The samples of DNA that Celera was using were kept in a freezer near Smith's office. When he wanted to get some human DNA, he removed a vial of frozen white blood cells or sperm from the freezer. The vials had coded labels. He would thaw the sample of cells or sperm, then mix the material with salt water, along with a little bit of detergent. A typical human cell looks like a fried egg, and the nucleus of the cell resembles the yolk. The detergent mixes the whites and the yolks—rather like scrambling an egg. As the cell falls apart, strands of DNA spill out in the salt water. The debris, the broken bits of the cell, fall to the bottom of the vial, leaving tangles of DNA suspended in the liquid.

One of Smith's research associates, a woman named Cindi Pfannkoch, showed me what shattered DNA was like. Using a pipette, she drew a tiny amount of liquid from a tube and let a drop fall to a sheet of wax, where it beaded up like a tiny jewel, the size of the dot over this *i*. An ant could have drunk it in full.

"There are two hundred million fragments of human DNA in this drop," she said. "We call that a DNA library."

She opened a plastic bottle, revealing a white fluff. "Here's some dried DNA." She took up a pair of tweezers and dragged out some of the fluff. It was a wad of dried DNA from the thymus gland of a calf. The wad was about the size of a cotton ball, and it contained several million miles of DNA.

"In theory," Ham Smith said, "you could rebuild the entire calf from any bit of that fluff."

I placed some of the DNA on the ends of my fingers and rubbed them together. The stuff was sticky. It began to dissolve on my skin. "It's melting—like cotton candy," I said.

"Sure. That's the sugar in DNA," Smith said.

"Would it taste sweet?"

"No. DNA is an acid, and it's got salts in it. Actually, I've never tasted it."

Later, I got some dried calf DNA. I placed a bit of the fluff on my tongue. It melted into a gluey ooze that stuck to the roof of my mouth in a blob. The blob felt slippery on my tongue, and the taste of pure DNA appeared. It had a soft taste, unsweet, rather bland, with a touch of acid and a hint of salt. Perhaps like the earth's primordial sea. It faded away.

The DNA came from five anonymous donors who had contributed their blood or semen for use in Celera's human genome project. The donors included both men and women, and a variety of ethnic groups. "I wouldn't be surprised if one of the donors is Craig," Ham Smith remarked.

CRAIG VENTER grew up in a working-class neighborhood on the east side of Millbrae, on the San Francisco peninsula. "My father was a CPA all his life, and my mother was very much a Donna Reed kind of mother," he said. "We were middle-class at a time when being middle-class really meant you had no money. It was a very big deal when my dad's income went past twelve thousand dollars a year. He was a Mormon who had been excommunicated for smoking and drinking coffee, and he was proud of it. His single strongest character trait was absolute honesty to a fault." Venter's father died at age fifty-nine of a heart attack.

They lived near the railroad tracks. One of his favorite childhood activities, he said, was to play chicken on the tracks. He and his friends would stand on the tracks when a locomotive was coming, and the last kid to jump out of the way was the winner. In high school, his grades mostly stank, but he excelled in science and shop. He also became a champion swimmer and broke his league's records in the backstroke. "I was essentially grounded throughout high school—I was always in trouble," he said. "I was disinclined to take tests." He got F's by default. His favorite high school teacher was Gordon Lish, who later became a distinguished novelist and editor. "Gordon Lish got fired from my high school for supposedly un-American activities," Venter said. "He slouched during the Pledge of Allegiance—he couldn't or wouldn't stand up straight. When he was fired for this, I led a demonstration that turned into a riot, and we shut down the school. The principal called me into his office and said, 'You must be getting extraordinary grades from this Lish.' I said, 'No, I'm getting F's, but I deserve them.' "

During his senior year of high school, Craig Venter spent a lot of time surfing in Half Moon Bay. After high school, he attended two junior colleges in a desultory way, but mostly he surfed. At that time, Venter had long blond hair and a beautiful body. Then he got a draft notice. He quickly enlisted in the Navy to avoid having to serve in the Army. ("My parents had both been in the Marine Corps, and they looked at the Army as the lowest form of life.") He ended up getting trained as a medical corpsman. He worked at the Navy hospital in San Diego, and ended up running a tuberculosis ward. He developed a passion for medicine. "Things clicked in for me, all of a sudden. I got hungry for knowledge," he said. He served as a medical corpsman in Vietnam, and twice he was sentenced to the brig for disobeying orders.

Venter had a history of confrontation with government authorities. As an enlisted man in San Diego, he was court-martialed for refusing a direct order given by an officer. "She happened to be a woman I was dating," Venter said. "We had a spat, and she ordered me to cut my hair. I refused." A friend of Venter's, Ron Nadel, who was a doctor in Vietnam, recalled that one of Venter's blowups with authority involved "telling a superior officer to do something that was anatomically impossible." Venter worked for a year in the intensive care ward at Da Nang hospital, where, he calculated, more than a thousand Vietnamese and American soldiers died during his shifts, many of them while the 1968 Tet Offensive was raging, the North Vietnamese and the Vietcong launching bloody attacks on American positions across South Vietnam. "That was when I learned that our government lies," Venter said. "I can't say how many thousands of cases and how many deaths occurred in the Da Nang hospital during the Tet Offensive. And I'd get letters back from friends in the United States saying that the newspapers were saying it wasn't that bad, there were only a few hundred casualties. The government was lying. I turned twenty-one in Vietnam." When he returned to the United States, Venter finished college, then earned a PhD in physiology and pharmacology from the University of California at San Diego. His discovery, in Vietnam, that the government lied seemed to be at the center of his relationship with the rival Human Genome Project. They were the government: they lied.

Venter got married to a molecular biologist, Claire Fraser, who was the president of The Institute for Genomic Research (TIGR, pronounced "Tiger"), in Rockville, a nonprofit institute that he and Fraser had helped establish in 1992. In 1998, he endowed TIGR with half of his original stake in Celera—five percent of the company. The money would be used to analyze the genomes of microbes that cause malaria and cholera and other diseases. (Venter and Fraser divorced in 2004.)

A few years ago, Venter developed a hole in his intestine, due to diverticulitis. He collapsed after giving a speech, and nearly died. He recovered, but he blamed the stress caused by his enemies for his burst intestine. Venter had enemies of the first order. They were brilliant, famous, articulate, and regularly angry at him. At times, Venter seemed to thrive on his enemies' indignation with an indifferent grace, like a surfer shooting a tubular wave, letting himself be propelled through their cresting wrath. At other times, he seemed baffled, and said he couldn't understand why they didn't like him.

One of Venter's most distinguished enemies, at the time, was James D. Watson, who, with Francis Crick and Maurice Wilkins, had won the Nobel Prize for discovering the shape of the DNA molecule—what they called the double helix. They did this work in 1953, and it changed forever the direction of biology. Their discovery showed that all the processes of life were encoded in a molecule, which, in theory, could be decoded—read like a book. James Watson helped found the

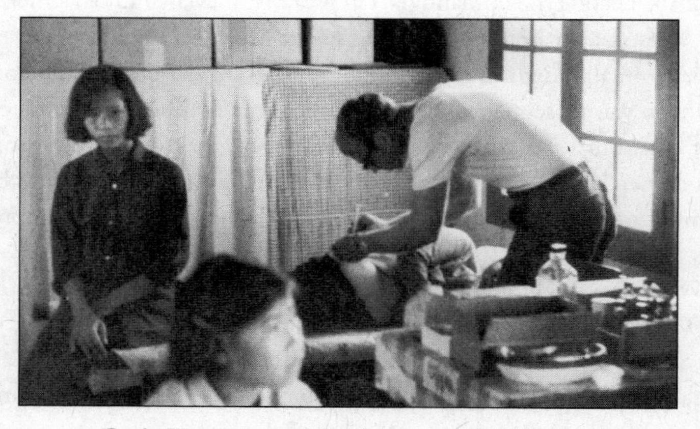

Craig Venter as a medical corpsman in Vietnam.

Courtesy of J. Craig Venter

Human Genome Project, and he was the first head of the NIH genome program. I visited him one day in his office at the Cold Spring Harbor Laboratory, on Long Island; he was the president of the laboratory. His office was paneled in blond oak, with a magnificent eastward view across Cold Spring Harbor. Watson was then in his seventies. He had a narrow face, lopsided teeth, a frizz of white hair, sharp, restless eyes, a squint, and a dreamy way of speaking in sentences that trailed off. He put his hands on his head and squinted at me. "In 1953, with our first paper on DNA, we never saw the possibility . . ." he said. He looked away, up at the walls, and didn't finish the sentence. "No chemist at the time ever thought we could read the molecule," he went on. But he, along with a number of biologists, began to think that reading the human DNA might just be possible. If the human book of life could be read, then the causes of many human diseases could be found and understood, and could be cured.

By the mid-1980s James Watson had become convinced that the decryption of the genome was an important goal and should be pursued, even if it cost billions and took decades. Part of his motivation might have been personal. James Watson had many eccentricities. He had a son who also seemed eccentric and, according to Watson, was not able to fully take care of himself. James Watson loved his son, and in witnessing his son's problems, it would be understandable that he ached to decode the human DNA in order to alleviate human suffering.

Watson appeared before Congress in May 1987 and asked for an initial annual budget of thirty million dollars for the project. The original plan was to sequence the human genome by 2005, at a projected cost of about three billion dollars. The principal work of the project was carried out by five major DNA-sequencing centers, as well as by a number of smaller centers around the world—all academic, nonprofit labs. The big centers included one at Baylor University, in Texas; one at Washington University, in St. Louis; the Whitehead Institute at MIT; the Joint Genome Institute of the Department of Energy; and the Sanger Centre, near Cambridge, England. The Wellcome Trust of Great Britain—the largest nonprofit medical research foundation in the world—funded the Sanger work, which was to sequence a third of the human genome. One of the founding principles of the Human

Genome Project was the immediate release of all the human code that was found, making it available free of charge and without any restrictions on who could use it or what anyone could do with it.

In 1984, Craig Venter had begun working at the NIH, where he eventually developed an unorthodox strategy for decoding bits of genes. At the time, other scientists were painstakingly reading the complete sequence of each gene they studied. This process seemed frustratingly slow to Venter. He began isolating what are called expressed sequence tags, or ESTs, which are fragments of DNA at the ends of genes. When the ESTs were isolated, they could be used to identify genes in a rough way. With the help of a few sequencing machines, Venter identified bits of thousands of human genes. This was a source of unease at the NIH, because it was a kind of skimming rather than a complete reading of genes. Venter published his method in 1991 in an article in *Science,* along with partial sequences from about 350 human genes. The method was not received well by many genomic scientists. It was fast, easy, and powerful, but it didn't look elegant; it looked like an application of brute force, and some scientists seemed threatened by it. Venter claimed that two of his colleagues, who eventually became heads of public genome centers, had asked him not to publish his method or move forward with it for fear they would lose their funding for genome sequencing done *their* way.

The NIH decided to apply for patents on the gene fragments Venter had identified. James Watson blew his stack over the idea of anyone trying to patent bits of genes. He got into a hostile situation with the director of the NIH, Bernadine Healy, who defended the patenting effort. In July 1991, during a meeting in Washington called by Senator Pete Domenici, of New Mexico, to review the genome program, Watson disparaged Craig Venter's methods. "It isn't science," he said, adding that Venter's machines "could be run by monkeys."

It was a strange moment. The Senate hearing room was almost empty—few politicians were interested in genes then. But Craig Venter was sitting in the room. "Jim Watson was clearly referring to Craig as a monkey in front of a U.S. senator," another scientist who was there said to me. "He portrayed Craig as the village idiot of genomics." Venter seemed to almost thrash in his chair, stung by Watson's words. "Watson was the ideal father figure of genomics," Venter says. "And he was attacking me in the Senate, when I was relatively young and new in the field."

That day in his office in Cold Spring Harbor, James Watson insisted to me that he hadn't been comparing Craig Venter to a monkey. "It's the patenting of genes I was objecting to. That's why I used the word 'monkey'! I hate it!" he said testily. The patent office turned down the NIH's application for a patent, anyway. But a few years later, two genomics companies, Incyte and Human Genome Sciences, adopted Craig Venter's EST method for finding genes, and it became the foundation of their businesses. Those businesses, combined, were worth many billions of dollars on the stock market. Samuel Broder, the chief medical officer at Celera, who was a former director of the National Cancer Institute, said to me, heatedly, "None of the people who severely and acrimoniously criticized Craig for his EST method ever said they were personally sorry. They ostracized Craig and then went on to use his method with never an acknowledgment."

James Watson said, "The EST method has proved immensely useful, and it should have been encouraged."

Venter was increasingly unhappy at the NIH. He had received a ten million dollar grant to sequence human DNA, and he asked for permission to use some of the money to do EST sequencing, but his request was denied by the Human Genome Project (which James Watson was then running). Venter returned the grant money with what he says was a scathing letter to Watson. In addition, Venter's wife, Claire Fraser, had been denied tenure at the NIH. Her review committee (which was composed entirely of middle-aged men, she said) explained to her that it could not evaluate her work independently of her husband's. At the time, Fraser and Venter had separate labs and separate research programs. Fraser considered suing the NIH for sex discrimination.

Meanwhile, James Watson, as head of the Human Genome Project at the NIH, had gotten himself into continuing scrapes with the head of the NIH, Bernadine Healy. During a press conference at which Healy was present, Watson criticized the NIH's policy of seeking patents on genes, and he labeled Healy (who was his boss) a "lunatic."

Shortly afterward, Bernadine Healy fired James Watson. She forced him to resign from his position as head of the Human Genome Project. Watson had done himself in with his mouth.

That summer, Craig Venter was approached by a venture capitalist named Wallace (Wally) Steinberg, who wanted to set up a company that would use Venter's EST method to discover genes, create new drugs, and make money. "I didn't want to run a company, I wanted to keep doing basic research," Venter said. But Wally Steinberg offered Venter a research budget of seventy million dollars over ten years—a huge amount of money, then, for biotech. Venter, along with Claire Fraser and a number of colleagues, left the NIH and founded TIGR, which is a nonprofit organization. At the same time, Wally Steinberg established a for-profit company, Human Genome Sciences, to exploit and commercialize the work of TIGR, which was required to license its discoveries exclusively to its sister company. Thus Venter got millions of dollars for research, but he had to hand his discoveries over to Human Genome Sciences for commercial development. Venter had one foot in the world of pure science and one foot in a bucket of money.

By 1994, the Human Genome Project was mapping the genomes of model organisms, which included the fruit fly, the roundworm, yeast, and E. coli (a bacteria that lives in the human gut), but no genome of any organism had been completed, except for virus genomes, which are relatively small. Venter and Hamilton Smith (the Nobel laureate and discoverer of DNA "scissor" enzymes, who was then at the Johns Hopkins School of Medicine) proposed speeding things up by using a technique known as whole-genome shotgun sequencing. In shotgunning, the genome is broken into small, random, overlapping pieces, and each piece is sequenced, or read. Then the jumble of pieces is reassembled in a computer that compares each piece to every other piece and matches the overlaps, thus assembling the whole genome. It's quite a lot like putting together a picture puzzle by matching the edges of the pieces to neighboring pieces, except that a genome can consist of millions of pieces and the task of matching them up to form a whole image of a genome requires superpowerful computers and really hot software.

Venter and Smith applied for a grant from the NIH to shotgun-sequence the genome of a disease-causing bacterium called *Haemophilus influenzae,* or *H. flu* for short. It causes fatal meningitis in children. They proposed to do it in just a year. *H. flu* has 1.8 million letters of code, which seemed massive then (though the human code is two thousand times as long). The review panel at the NIH gave Venter's proposal a low score, essentially rejecting it. According to Venter, the panel claimed that an attempt to shotgun-sequence a whole microbe was excessively risky and perhaps impossible. He appealed. The appeals process dragged on. While the appeals dragged on, he went about shotgunning *H. flu* anyway. Venter and the TIGR team had nearly finished sequencing the *H. flu* genome when, in early 1995, a letter arrived at TIGR saying that the appeals committee had denied the grant on the ground that the experiment wasn't feasible. Venter published the *H. flu* genome a few months later in *Science.* Whole-genome shotgunning had worked in spite of the objections of a funding committee at the NIH; it was almost as if the NIH wanted to *prevent* Venter's method from working. The method worked very well, however; this was the first completed genome of a free-living organism.

It seems quite possible that Venter's grant was denied because of politics. The NIH, the National Institutes of Health, is funded by tax dollars. The review panel seems to have hated the idea of giving taxpayer money to TIGR to make discoveries that would be turned over to a corporation, Human Genome Sciences, which could then profit from the discoveries. It turned down the grant in spite of the fact that "all the smart people knew the method was straightforward and would work," said Eric Lander, one of the leaders of the Human Genome Project.

Around this time, the venture capitalist Wally Steinberg died of a heart attack, and his death provided a catalyst for a split between TIGR and Human Genome Sciences, which was run by a former AIDS researcher, William Haseltine. Venter and Haseltine were widely known to despise each other. Venter had already sold his stock in Human Genome Sciences because of the rift between them, and after Steinberg died the relationship between the two organizations was formally ended.

Late in 1997, TIGR was doing some DNA sequencing for the Human Genome Project, and Venter began going to some of the project's meetings. That was when he started calling the heads of the public project's DNA-sequencing centers the Liars' Club, claiming that their predictions about when they would finish a task and how much it would cost were lies. His calling them liars did not win him many friends.

Francis Collins, a distinguished medical geneticist from the University of Michigan, had become the head of the NIH genome program shortly after James Watson resigned in 1992. In early January 1998, an internal budget projection from Collins's office somehow made its way to Watson (he seemed able to find out anything that was happening anywhere in molecular biology). This budget projection was supposed to be secret. It was an "eyes only" document intended to be seen by only about eight people, namely the top heads of the Human Genome Project. It contained a graph marked "Confidential" indicating that Francis Collins planned to spend only sixty million dollars per year on direct human-DNA sequencing through 2005. This was peanuts. It also predicted that by that year—when the human genome was supposed to be completed—actually only 1.6 billion to 1.9 billion letters of human code would be sequenced; that is, slightly more than half of the human genome would be done by then. The implication was that the whole human genome wouldn't get done until maybe 2008 or afterward.

This upset James Watson. Watson had hoped that his successor, Francis Collins, would aggressively pursue the human genome and get the work done as fast as possible; for one thing, he, James Watson, wanted to be alive to see the human genome. Second, the millions of people around the world who suffered from genetic diseases weren't getting any younger. He decided to discuss it with Eric Lander, the head of the Human Genome Project's DNA-sequencing center at MIT. Lander was a voluble, articulate, brilliant man in his forties who projected a high degree of self-confidence. He spoke in a rapid voice. He had a broad face and a mustache, and he owned stock in biotech companies that he advised or had helped to found. Eric Lander had become quite wealthy.

On January 17, James Watson traveled to Rockefeller University, on the East Side of Manhattan, where Eric Lander was giving the prestigious Harvey Lecture. The two men met after the lecture at the faculty club at Rockefeller. They were dressed in tuxedos, and Eric Lander had been drinking. By long tradition among medical people, the Harvey Lecture is given and listened to under the influence. Watson himself seemed a little tipsy. The scientists continued to drink after the lecture.

The Rockefeller faculty club overlooks a lawn and sycamore trees and the traffic of York Avenue. Watson and Lander sat down with cognacs at a small table in a dim corner of the room, on the far side of a pool table, where they could talk without being overheard. Also present and drinking cognac was a biologist named Norton Zinder, who was one of Watson's best friends. Zinder, like Watson, was a founder of the Human Genome Project. One of the older men brought up the confidential budget document with Lander, and both of them began to press him about it. They felt that it provided evidence that Francis Collins did not intend to spend more than sixty million dollars a year on human-DNA sequencing—and this was nowhere near enough money to get the job done anytime soon, they felt.

Watson evidently believed that Eric Lander had influence with Francis Collins, and he urged him to try to persuade Collins to spend more on direct sequencing of human DNA, and to twist Congress's arm for more money.

Norton Zinder was somewhat impaired with cocktails. "This thing is potchkying along, going nowhere!" he said, hammering the little table and waving his arms as he spoke. For him, the issue was simple: he had had a quadruple coronary bypass, and he had been receiving treatments for cancer, and now he was afraid he would not live to see the deciphering of the human genome. The human genome had begun to seem like a vision of Canaan to Norton Zinder, and he thought he wouldn't make it there.

Eric Lander did not view things the way the older biologists did. In his opinion, the problem was organizational. The Human Genome Project was "too bloody complicated, with too many groups." He felt the real problem was a lack of focus. He wanted the project to create a small, elite group that would do the major sequencing of human DNA—shock cavalry that would lead a charge into the human genome.

The three men downed their cognacs with a gloomy sense of frustration. "I had essentially given up on seeing the human genome in my lifetime," Zinder said.

NOT LONG BEFORE Watson and his friends began lamenting the slowness of the public project, the Perkin-Elmer Corporation, which was a manufacturer of lab instruments, had started secretly talking about an ambitious corporate reorganization. It controlled more than 90 percent of the market for DNA-sequencing equipment, and it was developing the Prism machine. The Prism was then only a prototype sitting in pieces in a laboratory in Foster City, California, but already it looked as if it was going to be at least ten times faster than any other DNA-sequencing machine. Perkin-Elmer executives began to wonder just what it could do. One day Michael Hunkapiller, who was then the head of the company's instrument division, got out a pocket calculator and estimated that several hundred Prisms could whip through a molecule of human DNA in a few weeks, although only in a rough way. To fill in the gaps—places where the DNA code came out garbled or wasn't read properly by the machines—it would be necessary to sequence the molecule again and again. This is known as repeat sequencing, or manyfold coverage, and he thought it might take a few years. Hunkapiller persuaded the chief executive of Perkin-Elmer, Tony White, to restructure the business and create a genomics company.

In December 1997, executives from Perkin-Elmer began telephoning Craig Venter to see if he'd be interested in running the new company. He blew them off at first, but a few months later he went to California with a colleague, Mark Adams, to check out the prototype Prism. When they saw it, they immediately understood its significance. They were looking at the equivalent of a supersonic jet in relation to a propellor aircraft. Before the end of that day, Venter, Adams, and Hunkapiller had laid out a plan for decoding the human genome *fast*. A month later, Norton Zinder, Watson's friend, flew to California to see the machine. Zinder saw it, too. "It was just a piece of equipment sitting on a table, but I said, 'That's it! We've got the genome!' " he recalled. Zinder joined Celera as a member of its board of advisers, and received stock in the company, which considerably enriched him. Now he could take a lesson from Eric Lander; he could cash in on the biotech boom. ("And what's wrong with that?" he asked me. "The chemists have been cleaning up all their life. Now the biologists are starting to get their hands on the money, and people are saying, 'Whoo, that's not kosher!' What's not kosher about it?")

After Norton Zinder got involved financially and scientifically in the Celera effort to race past the Human Genome Project, it led to some strains between him and James Watson. They maintained their friendship but finally had to agree not to speak about Celera with each other. They evidently feared that one or both of them could have a stroke arguing about Craig Venter.

One day not long after Norton Zinder saw the Prism machine and realized it was going to revolutionize the reading of DNA, Craig Venter and Mike Hunkapiller walked into the office of Harold Varmus, the director of the NIH, to talk to him about something. Harold Varmus was a Nobel laureate and an expert in genes and DNA. He had won the Nobel Prize in Medicine in 1989 (with J. Michael Bishop) for a theory of cancer-causing genes—a model of how cancer arises from genes embedded in a person's code. In Varmus's office that day, Craig Venter wanted to talk about the human code and the ongoing effort to read it. He announced the pending formation of a corporation, to be led by himself, that was going to decode the human genome. (Celera did not yet have a name.) Venter proposed to Varmus that the company and the public project collaborate, sharing their data and—this point is enormously important to scientists—sharing the publication of the human genome, which meant sharing the credit and the glory for having done the work. This included, of course, the unspoken possibility of the Nobel Prize in Medicine. The Nobel Prize would seem to have been *made* for the team that first decrypted the human DNA.

Harold Varmus was skeptical. He suspected that this wasn't a sincere offer from Craig Venter. He wondered if Venter might be angling for something that would be good for Craig Venter but maybe not so good for the Human Genome Project. He told Venter that he needed time to consider the proposal, particularly to check back with his subordinate Francis Collins (the head of the NIH's part of the Human Genome Project), to see what Collins thought of this unusual offer of collaboration.

Later that same day, Craig Venter and Mike Hunkapiller drove to Dulles Airport, where they met Francis Collins at the United Airlines Red Carpet Club. There they personally offered collaboration to Collins.

Venter recalled later that Collins seemed upset with his offer. Collins recalled that he merely asked Venter for some time to consider it. Extra time was one thing that Craig Venter was not prepared to give Francis Collins.

Venter was no stranger to ways of getting attention in the news media. By the time he met with Francis Collins, he had already alerted *The New York Times* to the creation of the new company to sequence human DNA. Just an hour or so after the meeting with Francis Collins he called the *Times* and told the paper it should go ahead and run the story. In the published account, Venter announced that he would sequence the human genome four years ahead of the public project. He would do it, he claimed, for less than a tenth of the projected cost of the public project—that is, he'd do it for less than $200 million, against the $3 billion–plus price tag of the Human Genome Project. The *Times* reporter, Nicholas Wade, implied that the Human Genome Project might not meet its goals and might be superfluous, now that Craig Venter and Celera had come along and were proposing to do the job much faster and much more cheaply—and at zero expense to the taxpayer. Certainly Francis Collins could not have been thrilled when he opened *The New York Times* the next day and read this. He hadn't yet even given Craig Venter a reply to his offer of collaboration.

By now, there was no stopping Venter. Four days later, on May 12, Venter and Hunkapiller went to the Cold Spring Harbor Laboratory—James Watson's institute—where a meeting of the heads of the Human Genome Project was taking place. Venter got up and told them, in effect, that they could just give up and stop working, since he was going to sequence the human genome *tout de suite*. Later that week, sitting beside Harold Varmus and Francis Collins at a press conference, Craig Venter looked out at a roomful of reporters and suggested that biology and society would be better off if the Human Genome Project stopped reading human DNA and moved forward to do the genome of . . . the mouse.

It was a fart in church of magnitude nine. Venter hadn't really intended to sound so offensive, but he had never been able to keep his mouth under control in a delicate situation. "The mouse is essential for interpreting the human genome," Venter tried to explain, but that didn't help.

In the words of one head of a sequencing center who was at the Cold Spring Harbor meeting, "Craig has a certain lack of social skills. He goes into that meeting thinking everyone is going to thank him for doing the human genome himself. The thing blew up into a huge explosion." The head of another center recalled, "Craig came up to me afterward, and he said, 'Ha, ha, I'm going to do the human genome. You should go do the mouse.' I said to him, 'You bastard. You *bastard*,' and I almost slugged him."

They felt that Venter was trying to stake out the human genome for himself as a financial asset while at the same time stealing the scientific credit. They felt that he was belittling their work, telling them to just go do the mouse.

Furthermore, Venter said that he would make the human genome available to the public but would charge customers who wanted to see Celera's analyzed data, and this made James Watson livid. He did not like the idea of having to pay money to Craig Venter for what he felt was the human heritage, which should be open to all for free. Watson did not deign to attend Venter's presentation—apparently he stayed up in his blond-paneled office and made telephone calls or fumed—but he appeared in the lobby, where he walked around and, in his strange, drifting voice, said to people, "He's Hitler. This should not be Munich." To Francis Collins he said, "Are you going to be Churchill or Chamberlain?"

Venter left the meeting soon afterward. Watson's remarks got back to him, of course. Venter didn't appreciate being called the Hitler of the human genome by the discoverer of the structure of DNA. Craig Venter and James Watson seemed to stop speaking with each other after that.

"You have to understand something about Jim Watson," Watson's friend Norton Zinder explained to me. "Jim has a kind of verbal Tourette's syndrome. He shoots his mouth off, and he doesn't know what he's saying. He can't control it." In this respect, Watson was remarkably like Craig Venter, Zinder pointed out. "Anyway, I wouldn't want to be Jim Watson," Zinder remarked.

"Why not?"

"Are you kidding? All he does is fly around the world to meetings, where he accepts another medal for something he did in 1953. It's a horrible life. I suppose he likes it."

The British leaders of the public project—John Sulston, the director of the Sanger Centre, and Michael Morgan, of the Wellcome Trust—reacted swiftly to Craig Venter's announcement. They were in England, but they flew to the United States and the next day arrived at Cold Spring Harbor, where they found things in disarray, if not total fibrillation, over Venter's announcement, with scientists wondering if the Human Genome Project was going to die. To a standing ovation, Michael Morgan got up and played the role of Winston Churchill. He read a statement declaring that the Wellcome Trust would nearly double its funding for the public project, and would challenge any "opportunistic" patents of the genome. "We were reacting, in part, to Craig's suggestion that we just close up shop and go home," Morgan later explained to me.

Venter also announced that Celera would use the whole-genome shotgun method—once again, as with his EST method, he was pushing the envelope of the possible, reaching for a new but seemingly risky technique to speed up the work of decoding the letters of DNA. The public project had chosen a more conventional method. John Sulston and Robert Waterston, the head of the sequencing center at Washington University, published a letter in *Science* asserting that Venter's method would be "woefully inadequate." Francis Collins was quoted in *USA Today* as saying that Celera was going to produce "the Cliffs Notes or the Mad Magazine version" of the human genome. (Collins later said that his words had been taken out of context by the reporter, and that he regretted the quote.) Norton Zinder, Watson's friend, told me that he wasn't at all surprised that Celera was getting ready to cream the government and decode the human DNA first. "The government will never be able to move as fast as a company," he said. "Anyway, it's an industrial job! That's why Celera is beating the crap out of the government."

THE COMPANY forged from Perkin-Elmer amid the turmoil was the PE Corporation, which was divided into two pieces, the PE Biosystems Group, the unit that was making the Prism machines, and Celera Genomics, which was using the machines to decode the human DNA. Michael Hunkapiller, who became the president of PE Biosystems, believed that he could sell a lot of machines to everyone, including to the Human Genome Project. Craig Venter's project would demonstrate how effective the Prism machines were; it was advertising. The deal was that there was a fat profit margin in the chemicals the machines used. The chemicals had a much higher profit margin than the machine; not only that, but the chemicals actually cost far more than the machine over the machine's lifetime. This was the razor-blade principle: if you put inexpensive razors in people's hands, you will make money selling blades.

James Watson quietly went to some key members of Congress and persuaded them to spend more money on the public project. At the same time, the leaders of the project announced a radical new game plan: they would produce a "working draft" of the human genome a year *ahead* of when Venter said he'd be done. An epic race had begun.

Michael Morgan, of the Wellcome Trust, told me what he thought had happened with the creation of Celera. "From the first press release, Craig saw the public program as something he wanted to denigrate," Morgan said. "This was our first sign that Celera was setting out to undermine the international effort. What is it that motivates Craig? I think he's motivated by the same things that drive other scientists—personal ego, a degree of altruism, a desire to push human knowledge forward—but there must be something else that drives the guy. I think

Craig has a huge chip on his shoulder that makes him want to be loved. I actually think Craig is desperate to win a Nobel Prize. He also wants to be very, very rich. There is a fundamental incompatibility there."

One day, I ran into a young player in the Human Genome Project. He believed in the worth and importance of the public project and said that he had turned down a job offer from Celera. He didn't have any illusions about human nature, or about any of the major players. He said, "Here's why everyone is so pissed at Craig. The whole project started when James Watson persuaded Congress to give him money for the human genome, and he turned around and gave it to his friends—they're the heads of centers today. It grew into a lot of money, and then the question was, Who was going to get the Nobel Prize? In the United States, there were seventeen centers in the project, and there was no quality control. It didn't matter how bad your data was, you just had to produce it, and people weren't being held accountable for the quality of their product. Then Celera appeared. Because of Celera, the NIH was suddenly forced to consolidate its funding. The NIH and Francis Collins began to dump more than eighty percent of the money into just three centers—Baylor, Washington University, and MIT—and they jacked everybody else. They had to do it, because they had to race Celera, and they couldn't control too many players. So all but three centers were cut drastically, and some of the labs closed down. Celera was not just threatening their funding but threatening their very lives and everything they had spent years building. It's kind of sad. Now those people hang around meetings, and the leaders treat them like 'If you're really nice, we'll give you a little piece of the mouse genome.' That's the reason so many of them are so angry at Celera. It's easier for them to go after Craig than to go after Francis Collins and the NIH."

AT CELERA'S HEADQUARTERS in Rockville, I was shown how human DNA was shotgunned into small pieces when it was sprayed through a hospital nebulizer that cost a dollar fifty. The DNA fragments were then introduced into E. coli bacteria and grown in glass dishes. The bacteria formed brown spots—clones—on the dishes. Each spot had a different fragment of human DNA growing in it. The dishes were carried to a room where three robots sat in glass chambers the size of small bedrooms. Each robot had an arm that moved back and forth rapidly over a dish. Little needles on the arms kept stabbing down and taking up the brown spots. Later, the bits of human DNA in the bacteria would be separated from the bacteria and run through the sequencing machines, producing little bits of human DNA code.

Craig Venter stood watching the robots move. The room smelled faintly like the contents of a human intestine. "This used to be done by hand," he said. All the human DNA fragments would eventually wind up in the Prism sequencing machines, and what would be left, at the end, was a collection of up to twenty-two million random fragments of sequenced human DNA. Then the river of shattered DNA would come to the supercomputer, and to a computer scientist named Eugene Myers, who with his team would assemble all the broken bits of human code into the more-or-less correct order, producing the full human genome.

GENE MYERS had dark hair and a chiseled, handsome face. He wore glasses, a green half-carat emerald in his left ear, and brown Doc Martens shoes. He also had a ruby and a sapphire that he would wear in his ear, instead of the emerald, depending on his mood. He was sensitive to cold. On the hottest days of summer, Myers wore a yellow Patagonia fleece jacket, and he kept a scarf wrapped around his neck. "My blood's thin," he explained to me. He said the scarf was a reference to the DNA of whatever organism he happened to be working on. When I first met Myers, he was keeping himself warm in his fruit-fly scarf. It had a black-and-white zigzag pattern. Later, Myers started wearing his human scarf, which had a green chenille weave of changing stripes. He intended his scarf to make a statement about the warfare between Celera and the public project. "I picked green for my human scarf because I've heard that green is a positive, healing color," he said. "I really want all this bickering to go away." His office was a cubicle in a sea of cubicles, most of which were stocked with Nerf guns, Stomp Rockets, and plastic Viking helmets. Occasionally, Myers would put "Ride of the Valkyries" on a boom box, and in a loud voice he would declare war. Nerf battles swept through Celera whenever the tension rose. Myers fielded a compound double-action Nerf Lock 'n Load Blaster equipped with a Hyper Sight. "Last week we slaughtered the chromosome team," he told me.

IN THE FALL, Venter announced that Celera had completed the sequencing of the fruit fly's DNA and had begun to run human DNA through its sequencing machines—there were now three hundred of them crammed into Building One in Rockville. The Command Center was up and running, and from then on Celera operated in high-speed mode. One day that fall, I talked with the company's information expert, a stocky man named Marshall Peterson. He took me to the computer room, in Building Two. To get into the room, Peterson punched in a security code and then placed his hand on a sensor, which read the unique pattern of his palm. There was a clack of bolts sliding back. We pushed through the door.

A chill of cold air washed over us, and we entered a room filled with racks of computers that were wired together. "We have fifty-five miles of fiber-optic cables running through this building," Peterson said. Workmen standing on ladders were installing many more cables in the ceiling. "The disk storage in this room is five times the size of the Library of Congress. We're getting more storage all the time. We need it."

He took me to the Command Center, where a couple of people were hanging around consoles. A big screen on the wall showed CNN Headline News. "I've got a full-time hacker working for me to prevent security breaches," Peterson said. "We're getting feelers over the Internet all the time—people trying to break into our system." Celera would be dealing with potentially valuable information about the genes of all kinds of organisms. Peterson thought that some of what he called feelers—subtle hacks and unfriendly probes—had been emanating from Celera's competitors. He said he could never prove it, though. Lately, the probes had been coming from computers in Japan. He thought it was American hackers co-opting the Japanese machines over the Internet.

By October 20, forty days after Celera started running human DNA through its machines, the company announced that it had sequenced 1.2 billion letters of human code. The letters came in small chunks from all over the genome. Six days later, Venter announced that Celera had filed provisional patent applications for sixty-five hundred human genes. The applications were for placeholder patents. The company hoped to figure out later which of the genes would be worth patenting in earnest.

A gene patent gives its holder the right to make commercial products and drugs derived from the gene for a period of seventeen years. Pharmaceutical companies argue that patents are necessary, because without them businesses would never invest the hundreds of millions of dollars needed to develop a new drug and get it through the licensing process of the Food and Drug Administration. ("If you have a disease, you'd better hope someone patents the gene for it," Venter said to me.) On the other hand, parceling out genes to various private companies could lead to what Francis Collins referred to as the "Balkanization of the human genome," a paralyzing situation that might limit researchers' access to genes.

Venter insisted that Celera was an information company and that patenting genes was not its main goal. He had said that Celera would attempt to get patents on not more than about three hundred human genes. Even so, it was pretty clear that Celera was hoping to nail down some very valuable real estate in the human genome—billion-dollar genes, perhaps.

All summer long, Celera's stock had bounced around between seven and ten dollars a share. Around Halloween, as investors began to realize that the company was cranking out the human genome—and filing large numbers of patents on genes—the stock jumped up to twenty dollars a share. On December 2, the Human Genome Project announced that it had deciphered most of the code on chromosome no. 22, the second-shortest chromosome in the human genome. This made the reading of the whole genome seem doable and imminent, and Celera's stock began a spectacular, tornadic rise of a sort that has rarely been seen in the American stock market. It shot up that day by nine points, and closed at over seventy dollars. Then, after the market's close on Thursday, December 16, Tom Gardner, a cofounder of the Web site called the Motley Fool, announced that he was buying shares of Celera for his own portfolio. It was known as the Rule Breaker Portfolio, and it featured small companies that broke the rules and changed the landscape of business.

Celera came of age during the huge rise of the Internet stock-market bubble. When the news broke that Celera had been named to the Motley Fool's Rule Breaker Portfolio, a large number of people tried to buy Celera the next morning. They drove the stock up twenty points in a matter of minutes. A few months earlier, it had been trading at seven dollars a share. Celera's stock price looked like it was headed for Mars.

I went to visit Celera one day the following week. On that particular morning the company's stock could not even open for trading on the New York Stock Exchange. That morning, it seemed as though all of Wall Street wanted to buy Celera. That morning a tsunami of buy orders for Celera overwhelmed the specialists on the floor of the New York Stock Exchange. Trading in Celera froze, while the traders on the floor of the New York Stock Exchange waited for sell orders to trickle in. While the stock was halted—at $101 a share—I wandered around the building.

There was a feeling of shock in the air. Everyone was aware of the trading halt in the stock; everyone in the building owned Celera stock. Just about every employee of Celera was becoming a multimillionaire, and it seemed to be happening by the minute. I felt that very little work was getting done that day at Celera, except by the robots. Employees were checking the stock quote on Yahoo! and wondering what their net worth would be in an hour or two, when the stock would finally open and start trading.

I found Hamilton Smith in his lab, puttering around with human DNA in tiny test tubes. He seemed to be the only person at the company who wasn't very affected by the situation. He was tired and looked sleep-deprived. He explained that he was renovating his house and had stayed up all night ripping carpet out of the basement. "The carpet guys were coming in to lay new carpet in the basement, and I didn't feel like paying them to rip out the old carpet," he said. "It would have been expensive." Hamilton Smith owned many thousands of shares of Celera, and his net worth was already in the many millions. He also refused to buy a new car. He had driven to work that day, as usual, in his '83 Mercury Marquis.

Smith passed a computer, stopped, and brought up a quote. Celera had finally opened for trading. It had gapped upward—that is, it had jumped instantly upward—by thirteen points. It was at 114. Smith's net worth had jumped upward by around a million dollars in ten seconds. "Is there no end to this?" he muttered.

Craig Venter came into Smith's lab and asked him to lunch. In the elevator, Smith said to him, "I can't stand it, Craig. The bubble will break." They sat down beside each other in the cafeteria and ate cassoulet from bowls on trays.

"This defies common sense," Smith went on. "It's really impossible to put a value on this company."

"That's what we've been telling the analysts," Venter said.

Later that day, I ended up in Claire Fraser's office at TIGR headquarters, a complex of semi–Mission style buildings a couple of miles from Celera's offices and labs. Fraser, who was then Venter's wife, was a tall, reserved woman with dark hair and brown eyes, and her voice had a New England accent. She grew up in Massachusetts. In high school, she said, she was considered a science geek. Her office had an Oriental rug on the floor and a table surrounded by Chippendale chairs. ("This is Craig's extravagant taste, not mine," she explained.) Two poodles, Shadow and Marley, slept by a fireplace.

"Before genomics, every living organism was a black box," she said. "When you sequence a genome, it's like walking into a dark room and turning on a light. You see entirely new things everywhere."

Fraser placed a sheet of paper on the table. It contained an impossibly complicated diagram that looked like a design for an oil refinery. She explained that it was an analysis of the genome of cholera, a single-celled microbe that causes murderous diarrhea; TIGR scientists had finished sequencing the organism's DNA a few weeks earlier. Much of the picture, she said, was absolutely new to our knowledge of life. About a quarter of the genes of every microbe that had been decoded by TIGR were completely new to science and were not obviously related to any other gene in any other microbe. To the intense surprise and wonder of the scientists, nature was turning out to be an uncharted sea of unknown genes. The code of life was far richer and more beautiful than anyone had imagined.

Fraser's eyes moved quickly over the diagram. In effect, she was seeing cholera for the first time in the history of biology. And she could sight-read the diagram, in the same way that a good musician can sight-read Mozart and hear it in her head. "Yes . . . wow . . . ," Fraser murmured. "Wow. There may be important transporters here. . . . It looks like there could be potential for designing a new drug that could block them."

The phone rang. Fraser walked across her office, picked up the receiver, and said softly, "Craig? Hello. What? It closed at a hundred and twenty-five?" Pause. "I don't know how much it's worth. You're the one with the calculator."

Their net worth had jumped above $150 million that day.

Fraser drove home, and I followed her in my car. The house she shared with Venter was in a wealthy neighborhood outside Washington. It sat behind a security gate at the end of a long driveway. Venter arrived in a brand-new Porsche. The car would do zero to sixty in five seconds, he said. In the vaulted front hall of the house there was a model of HMS *Victory* in a glass case. In a room next to the garage, there was a jumble of woodworking machines—a band saw, a table saw, a drill press. Venter had worked with wood since his shop classes in high school.

In the kitchen, Claire fixed dinner for the poodles, while Craig circled the room, talking. "We created close to two hundred millionaires in the company today. I think most of them had not a clue this would happen when they joined Celera. We have a secretary who became a millionaire today. She's married to a retired policeman. He went out looking to buy a farm." He popped a Bud Light and swigged it. "This could only happen in America. You've got to love this country." Claire fed the poodles.

There were no cooking tools in the kitchen that I could see. The counters were empty. The only food I noticed was a giant sack of dog food, sitting on top of an island counter, and two boxes of cold cereal—Quaker Oatmeal Squares and Total. In the guest bathroom, upstairs, there were no towels, and the walls were empty. The only decorative object in the bathroom was a cheap wicker basket piled with little soaps and shampoos they had picked up in hotels.

We went to a restaurant and ate steak. "We're in the Wild West of genomics," Venter said. "Celera is more than a scientific experiment; it's a business experiment. Our stock-market capitalization as of today is three and a half billion dollars. That's more than the projected cost of the Human Genome Project. I guess that's saying something. The combined market value of the Big Three genomics companies—Celera, Human Genome Sciences, and Incyte—was about twelve billion dollars at the end of today. This wasn't imaginable six months ago. The Old Guard doesn't have control of genomics anymore." He chewed steak and looked at his wife. "What the hell are we going to do with all this money? I could play around with boats . . ."

Claire started laughing. "My God, I couldn't live with you."

"The money's nice, but it's not the motivation," Venter said to me. "The motivation is sheer curiosity."

In December, Celera and the Human Genome Project discussed whether it would be possible to collaborate. There was one formal meeting, and there were many points of difference. Meanwhile, Celera's stock seemed to go into escape velocity. In January, it soared to over $200 a share. Celera filed to offer more shares to the public and declared a two-for-one stock split. Shortly after the split, the stock hit an all-time high of $276 a share (adjusted for the split). Celera's stock-market value reached $20 billion, and Craig Venter's wealth on paper surpassed $1 billion. It looked as though Craig Venter had become the first billionaire of biotechnology.

Francis Collins, the director of the National Institutes of Health's part of the Human Genome Project—in actuality, he was the principal leader of the project—was a Christian who drove to work on a motorcycle. He played guitar in an amateur rock band. He was six feet four inches tall. He had expressive hands that moved while he spoke, and he had a soft, expressive voice. He wore faded jeans and motocycle boots or sneakers. His net worth was not impressive, because his employer was the government. He had a mop of graying brown hair combed over his forehead. Collins had grown up on a small farm in the Shenandoah Valley, where he'd been homeschooled by his mother, Margaret Collins, a noted American playwright, and by his father, Fletcher, a medieval scholar and founder of several theater troupes. Francis Collins went to Yale, where he got a PhD in quantum chemistry, but then he decided to become a medical doctor.

"My epiphany came two months into medical school, when I listened to a series of lectures on genetic diseases by an austere and impressive pediatric geneticist," Collins told me in his office on the campus of the National Institutes of Health, in Bethesda, Maryland. "Each week he brought a different child in front of the class, with a different genetic disease—cystic fibrosis, sickle-cell anemia. They're sick, I thought, because they have a single little thing wrong in the arrangement of their code. It was almost unbelievable that such a small change could have such large effects in some people. It was terrifying."

"Have you heard of Lesch-Nyhan syndrome?" I asked Collins, referring to the self-cannibalism disease. It is caused by a tiny defect in the human DNA, typically the alteration of a single letter of the human genome.

Francis Collins had certainly heard of the disease. He had seen it. "I diagnosed a case of Lesch-Nyhan syndrome when I was at Yale medical school," Collins said. "He was a young man, twenty-four years old, and he was engaging in self-mutiliation. I've thought a lot about Lesch-Nyhan syndrome. Why would a loving God allow the kind of suffering we see in a Lesch-Nyhan person? Why would God permit a loss of free will in these people? These are tough questions. We tend to see our suffering as a consequence of our own free-will choices. But when the suffering comes from a little glitch in our DNA, how is that compatible with a loving God? I can't say that I have the answers. Perhaps there is an answer in John, chapter nine, when Christ and his disciples pass by a child who is blind. One of the disciples asks, 'Master, who sinned that this child was born blind? Was it the child or his parents?' Christ answered, 'Neither have sinned. He is blind so that the nature of God might be manifest in him.' We don't learn much when life is too easy; God shouts at us in the tough times. Even so, I'm not sure that Christ's answer makes me comfortable with a child with a genetic disease. I've spent way too much time in hospitals with families and patients, putting names to diseases that I couldn't do a thing about. Who knows what lurks in our DNA?"

Francis Collins stretched out his long frame, sticking his legs out straight, and crossed his sneakers. "Sequencing a gene is one thing, but reading the whole human genome—that could never have been imagined. Between splitting the atom, and going to the moon, and reading the human DNA, I would argue that this last will go down in history as the most important. This simple code, with its amazing properties, which unifies all living forms . . ." His voice trailed off. "It carries me away."

Right around the time I spoke with Francis Collins, newspapers reported that the discussions between Celera and the public project had collapsed. It seems that the discussions had been going on via the two main principals, Craig Venter and Francis Collins. The toughest point of disagreement, according to officials at the public project, was that Celera wanted to keep control of intellectual property in the human genome, while Francis Collins and the other leaders of the Human Genome Project were determined to let the information be available to anyone for free. Celera's stock began to drop.

Then it went into a screaming nosedive. It dropped fifty-seven dollars in a matter of hours, lurching downward. On the floor of the New York Stock Exchange, the traders were holding fistfuls of sell slips for Celera, and nobody wanted to buy the stock. The other genomic stocks crashed in sympathy with Celera. This, in turn, dragged down the NASDAQ, which that day suffered the second-largest point loss in its history.

It was, in fact, a more generalized heart attack affecting the entire American stock market. This was the popping of the Internet stock-market bubble. It ushered in a bear market, a period of years in which the stock market went down or sideways. Not merely Internet stocks were crashing; everything that had anything to do with technology was falling in value. In the end, just about anything that had to do with the Internet was crushed, and many stocks lost 90 to 100 percent of their value. Short sellers—people who profit from the decline of a stock—had encrusted Celera's stock like locusts. Craig Venter seemed to be one of the forces bringing the stock market down.

"I feel a little like Galileo. I'm expecting a call from the pope any day now, asking me to recant the human genome," Venter said to me right after Celera's stock had gone over the lip of Niagara Falls in a barrel. He sounded wired and exhausted. "They offered to have a barbecue with Galileo, right? Look, I'm not likening myself to Galileo in terms of genius, but it is clear that the human genome is the science event of our time. I am going to publish the human genome, and that's what the threat to the public order is. Our publishing the genome makes a mockery of the fifteen years and billions of dollars the public project has spent on it."

Venter seemed particularly upset with the British part of the public project. "In my opinion," he said, "the Wellcome Trust is now trying to justify how, as a private charity, it gave what I think was well over a billion dollars to do just a third of the human genome, largely at the expense of the rest of British medical science. Forty billion dollars was taken out of the biotechnology industry today—that's how much was lost by investors in all the biotechnology companies. It was money that would pay for cures for cancer, and it was taken off the table, all because some bastards at the Wellcome Trust are trying to cover up their losses."

I called Michael Morgan, at the Wellcome Trust, to see what he had to say about this. "In hindsight, it is easy to ascribe to us Machiavellian powers that the prince would have been proud of," he said dryly. "As for the allegation that I'm a bastard, I can easily disprove it using the technology of the Human Genome Project."

"ONE OF THE THINGS I really like about Craig Venter is that he almost totally lacks tact," one of his collaborators, a genomic scientist from U.C. Berkeley, Gerald Rubin, told me. "If he thinks you are an idiot, he will say so. I find that way of dealing very enjoyable. Craig is like somebody who's using the wrong fork at a fancy dinner. He'll tell you what he thinks of the food, but he won't even think about what fork he's using. It was a great collaboration."

John Sulston, the head of the Sanger Centre, told the BBC that he felt Celera planned to try to get a monopoly on the human DNA. "The emerging truth is absolutely extraordinary," Sulston said. "They really do intend to establish a complete monopoly position on the human genome for a period of at least five years." He added, "It's something of a con job."

"Sulston essentially called us a fraud. It's like he's been bit by a rabid animal," Venter fumed.

"It's puzzling. To me, the whole fight defies rational analysis," Nobel laureate Hamilton Smith said to me, shortly after his net worth had cratered in Celera's mud slide. (He had continued to drive his '83 Marquis, so his lifestyle had not been much affected.) "But the publicly funded labs are angry for reasons I can partly understand," Smith went on. "We took it away from them. We took the big prize away from them, when they thought they would be the team that would do the whole human genome and go down in history. Pure and simple, they hate us."

AS FOR THE SCIENCE, most biologists seemed enthralled with the sight of the human DNA being decoded and revealed. They felt that a great door was opening and light was shining deep into nature, suggesting the presence of rooms upon rooms that had never been seen before. There was also a clear sense that the door would not have been opened so soon if Craig Venter and Celera had not given it a swift kick.

"We can thank Venter in retrospect," James Watson said, leaning back and smiling and squinting at the ceiling. "I was worried he could do it, and that would stop public funding of the Human Genome Project . . . but if an earthquake suddenly rattled through Rockville and destroyed Celera's computers, it wouldn't make much difference. . . ." His voice trailed off. He stood up and offered me the door.

Eric Lander, who said he liked Craig Venter personally, told me, "Having the human genome is like having a Landsat map of the earth, compared to a world where the map tapers off into the unknown with the words 'There be dragons.' It's as different a view of human biology as a map of the earth in the fourteen hundreds was compared to a view from space today." As for the war between Celera and the Human Genome Project, he said it was silly. "At a certain level, it was just boys behaving badly. It happened to be the most important project in science of our time, and it had all the character of a school-yard brawl."

ON FEBRUARY 16, 2001, close to three billion letters of the human DNA in their roughly proper order were published by Craig Venter (along with two hundred coauthors) in *Science* magazine, under the title "The Sequence of the Human Genome." It was a very good draft of the human genome, but it was not finished or fully accurate. That same week, the Human Genome Project published its own draft of the human genome in *Nature*. Taken together, these publications represented one of the monumental achievements of the twenty-first century. The implications would be with us for the rest our lives, the rest of our grandchildrens' lives, and their grandchildrens' lives. "Humanity has been given a great gift," declared an editorial in *Science*. "This stunning achievement has been portrayed—often unfairly—as a competition between two ventures, one public and one private. That characterization detracts from the awesome accomplishment unveiled this week."

Norton Zinder, Watson's friend, who had feared that he would die before he saw the human genome, said that he felt marvelous. "I made it. Now I've gotta stay alive for four more years, or I won't get all my options in Celera." Zinder was convinced that Celera's stock would rise again. (It has not recovered, so far.)

Norton Zinder was a vigorous-seeming older man. As he spoke, he was sprawled in a chair in his office overlooking the East River, gesturing with both hands raised. He shifted gears and began to look into the future. "This is the beginning of the beginning," he said. "The human genome alone doesn't tell you crap. This is like Vesalius. Vesalius did the first human anatomy." Vesalius published his work in 1543, an anatomy based on his dissections of cadavers. "Before Vesalius," Zinder went on, "people didn't even know they had hearts and lungs. With the human genome, we finally know what's there, but we still have to figure out how it all works. Having the human genome is like having a copy of the Talmud but not knowing how to read Aramaic."

CRAIG VENTER ended up getting fired from Celera. The man who fired him was Tony L. White, the chief executive of Celera's parent company (which had been renamed Applera). Tony White had been Craig Venter's boss during the time Venter worked at Celera—even though Venter was the president of Celera, Applera owned Celera, and so was in control. It seemed that Craig Venter's business model just wasn't working out. Venter had figured that Celera would sell information about the human genome to subscribers, who would want to pay money for it. But not enough customers wanted to pay for it.

The problem for Celera was that the Human Genome Project kept on churning out human DNA code and, with taxpayer money, published an increasingly accurate sequence of the letters in the human DNA—published it all for free on the Internet, available to anybody at no cost. Celera couldn't compete.

Craig Venter lost his job at the moment when Celera's stock was already crashing, and his firing from Celera caused the stock go into free fall, headed for near zero, it seemed. Venter had to sell all of his shares in Celera upon his exit from the company, and he sold them at the very bottom. He ended up walking away with about one million dollars in his pocket after his adventure with Celera. "I made a million dollars the hard way," he remarked. "I started with a billion dollars and lost it."

AFTER HE LEFT CELERA, Craig Venter went sailing. He still had some assets left, including shares in a company called Diversa, which he had cofounded years earlier. He had also sold *Sorcerer* (the yacht in which he'd won the Transatlantic Challenge) and he had gotten a few million dollars from that. He took some of the money and bought a used sailing yacht, a ninety-four-foot sloop, named it *Sorcerer II,* and outfitted it as a marine research vessel. He raised federal and private funding, hired a crew, and went on a voyage to circumnavigate the earth, following the idea of the HMS *Challenger* expedition of the 1870s, when the British sailing ship went on the first oceanographic expedition to explore the depths of the earth's oceans. "I was telling people that all I hoped to get out of this deal was a bigger yacht," he said.

The *Sorcerer II* sailed thirty-three thousand miles. Every two hundred miles, Venter and his colleagues stopped the boat and took samples of seawater. These bottles of water were sent back to the laboratory of a nonprofit foundation called the J. Craig Venter Institute, in Rockville, which Venter had established and ran.

Seawater is loaded with viruses, bacteria, and single-celled organisms. Therefore it's a soup with DNA floating around in it. The samples of seawater were run through DNA-sequencing machines, and the resulting fragments of DNA code were assembled in a supercomputer that looked for patterns. Venter was probing the sea, the heart of the earth's ecosystems, for undiscovered genes. To the great surprise of biologists, Venter's expedition revealed that the oceans are filled with millions of distinctly different and previously unknown species of microorganisms. The sea contained a vast, almost incalculable number of unknown life-forms, invisible to the eye. Especially viruses. The oceans of the earth appeared to contain perhaps a hundred million different kinds of viruses, virtually none of which were known to science, as well as countless millions of species of bacteria and single-celled organisms that had never been seen before.

Meanwhile, Craig Venter and Claire Fraser, who was the head of TIGR, had a divergence in their love lives and got divorced. Fraser left TIGR and moved to the University of Maryland, and eventually TIGR was merged into the J. Craig Venter Institute. This institute became a respected force in genomics. It had five hundred employees. Many of them, including Hamilton Smith, had exited from Celera en masse to follow Venter wherever he went.

Craig Venter had never given up sequencing the human genome. In the fall of 2007, he published a superaccurate and complete version of his own genome. Craig Venter's genome had both sets of chromosomes in it (humans have a double set of chromosomes in each cell).

Craig Venter's yacht Sorcerer II,
sailing to windward in search of unknown genes.
Courtesy of the J. Craig Venter Institute

This was a so-called diploid human genome, and it had more than six billion letters in it. It was twice as long as the Human Genome Project's human genome. Craig Venter's genome was the most finished, accurate, complete image of the human DNA that will likely be published.

He made his entire DNA code available for free on the Internet, as well. Thus anyone who wanted to know how Craig Venter, as an example of the human species, was constructed could read his blueprint on the Web site. The government may have thought it was only cruising, but Venter had never stopped racing.

CRAIG VENTER'S latest project was to create life in a test tube. He called it synthetic biology. The leader of the synthetic biology effort at at the J. Craig Venter Institute was Hamilton Smith. Smith, who had made and lost millions in the Celera venture, had finally sold his '83 Mercury Marquis and replaced it with a white Cadillac. Now Smith, Venter, and their colleagues were creating artificial genes—making stretches of DNA in the laboratory, rather like typing words with a typewriter. They were putting the machine-made DNA into bacteria. Their goal was to create synthetic organisms, microbes made by man to serve man. These laboratory-made bacteria, Venter and Smith hoped, would be able to digest cellulose plant material, such as in grass and pulpwood, ultimately turning it into ethanol, which could be used to power automobiles in place of gasoline.

Venter had founded a private company—for profit—called Synthetic Genomics. "We have some major deals with companies, such as one with BP [British Petroleum] to develop synthetic bacteria that would metabolize coal and turn it into natural gas," he said. "We're developing bacteria that can convert carbon dioxide that's been sequestered from the burning of coal and turn it into methane [for use in natural gas]. We've engineered a cell line"—a strain of living cells—"that could produce a new biological fuel that would replace jet fuel," he said. In this way, Craig Venter and Hamilton Smith hoped to help reduce global climate change while getting rid of America's dependence on foreign oil, all made possible by advances in the reading and writing of DNA. Along the way, Venter and his colleagues decoded the DNA of the dog. The dog DNA came from Shadow, one of Craig Venter and Claire Fraser's poodles. The last I heard, Shadow was living with Claire and doing fine. "I have visitation privileges with Shadow but see him rarely," Venter said.

The Lost Unicorn

In 1998, the Cloisters—the museum of medieval art in upper Manhattan—began a renovation of the room where the seven tapestries known as *The Hunt of the Unicorn* hang. The Unicorn Tapestries are considered by many to be the most beautiful tapestries in existence. They are also among the great works of art of any kind. In the tapestries, richly dressed noblemen, accompanied by hunters and hounds, pursue a unicorn through forested landscapes. They find the animal, appear to kill it, and bring it back to a castle; in the last and most famous panel, "The Unicorn in Captivity," the unicorn is shown bloody but alive, chained to a tree surrounded by a circular fence, in a field of flowers. The tapestries are twelve feet tall and up to fourteen feet wide (except for one, which is in fragments). They were woven from threads of dyed wool and silk—some of them gilded or wrapped in silver—around 1500, probably in Brussels or Liège, for an unknown person or persons, and for an unknown reason—possibly to honor a wedding. A monogram made from the letters *A* and *E* is woven into the scenery in many places; no one knows what it stands for. The tapestries' meaning is mysterious; the unicorn was a symbol of many things in the Middle Ages: Christianity, immortality, wisdom, lovers, marriage.

For centuries, the Unicorn Tapestries were in the possession of the La Rochefoucauld family of France; they hung in the family's château in Verteuil, a town in Charente, north of Bordeaux. François, the sixth duc de La Rochefoucauld (1613–1680) was a famed author of maxims. Among them is this one: "There is something in the misfortune of our friends that does not displease us."

In 1789, during the French Revolution, a mob of peasants looted the La Rochefoucauld château while the family fled. The family eventually returned to the château, but the Unicorn Tapestries had disappeared. Two generations later, in 1855, the then duc de La Rochefoucauld sent word around town that he'd like to buy the Unicorn Tapestries from anyone who might still have them. Some local farming people came forward and sold the tapestries back to the family. It turned out that the farmers had been using the Unicorn Tapestries to cover heaps of potatoes inside barns and to wrap fruit trees during the winter to keep them from freezing. Nevertheless, the tapestries were still in good shape. In the early 1920s, the La Rochefoucauld family, needing money, put the tapestries on display with an art dealer in Paris, and in 1922, John D. Rockefeller, Jr., bought them for just over a million dollars. He kept them in his apartment on Fifth Avenue, and in 1937 he gave them to the Cloisters. Their monetary value today is incalculable.

As the construction work at the Cloisters got under way, the tapestries were rolled up and moved, in an unmarked vehicle and under conditions of high security, to the Metropolitan Museum of Art, which owns the Cloisters. They ended up in a windowless room in the museum's textile department for cleaning and repair. The room has white walls and a white tiled floor with a drain running along one side. It is exceedingly clean and looks like an operating room. It is known as the wet lab, and is situated on a basement level below the museum's central staircase.

In the wet lab, a team of textile conservators led by a woman named Kathrin Colburn unpacked the tapestries and spread them out facedown on a large table, one by one. At some point, the backs of the tapestries had been covered with linen. The backings, which protected the tapestries and helped to support them when they hung on a wall, were turning brown and brittle, and had to be replaced. Using tweezers and magnifying lenses, Colburn and her team delicately removed the threads that held each backing in place. As the conservators lifted the backing away, inch by inch, they felt a growing sense of awe. The backs were almost perfect mirror images of the fronts, but the colors were different. Compared with the fronts, they were unfaded: incredibly bright, rich, and deep, more subtle and natural-looking. The backs of the tapestries had been exposed to very little sunlight in five hundred years—even, apparently, during the time when they had been used to cover potatoes. Nobody alive at the Met, it seems, had seen the backs of the Unicorn Tapestries in all their richest color.

A tapestry is woven from lengths of colored thread called the weft, which are passed around long, straight, strong threads called the warp. The warp runs horizontally and provides a foundation for the delicate weft, which runs vertically. Medieval tapestry weavers worked side by side, in teams, using their fingertips and small tools to draw the delicate weft threads around the tougher warp. When they switched from one color to the next, they cut off the ends of the weft threads or wove them into the surface of the tapestry. The Unicorn weavers had been compulsively neat. In less well-made tapestries, weavers left weft threads dangling on the back of the tapestry in a shaggy sort of mess, but the backs of these were almost smooth. Kathrin Colburn recalled that as she and her associates stared into the backs of the Unicorn Tapestries, it "felt like a great exploration of the piece." She said, "We simply got carried away, seeing how the materials were used—how beautifully they were dyed and prepared for weaving." An expert medieval weaver might have needed an hour to complete one square inch of a tapestry, which meant that in a good week he might have finished a patch maybe eight inches on a side. The weavers were generally young men, and each of the Unicorn Tapestries had likely had a team of between four and six working on it. They wove only by daylight, to ensure that the colors would be consistent and not distorted by candlelight. One tapestry would have taken a team at least a year to complete.

The curator in charge of medieval art at the Metropolitan and the Cloisters is a thoughtful man named Peter Barnet. When he heard about the discovery, he hurried down to the wet lab for a look. He got a shock. "The first of the tapestries—'The Start of the Hunt'—was lying in a clear, shallow pool of water," Barnet said. The lab is designed to function as a big tub, and had been filled about six inches deep with purified water to bathe the tapestry. "Intellectually, I knew the colors wouldn't bleed, but the anxiety of seeing a Unicorn tapestry underwater is something I'll never forget," he said. When Barnet looked at the

A modern tapestry weaver working on a tapestry.
She is able to finish around a square inch in an hour.
Richard Preston

image through the water, he recalled, "the tapestry seemed to be liquefied." Once the room had been drained, it smelled like a wet sweater.

Philippe de Montebello, the director of the museum, declared that the Unicorn Tapestries must be photographed on both sides, to preserve a record of the colors and the mirror images. Colburn and her associates would soon put new backing material on them, made of cotton sateen. Once they were rehung at the Cloisters, it might be a century or more before the true colors of the tapestries would be seen again. The manager of the photography studio at the Met was a pleasant, lively woman named Barbara Bridgers. Her goal was to make a high-resolution digital image of every work of art in the Met's collections. The job would take at least twenty-five years; there are roughly two and a half million cataloged objects in the Met—nobody knows the exact number. (One difficulty is that there seems to be an endless quantity of scarab beetles from Egypt.) But when it's done and backup files are stored in an image repository somewhere else, if an asteroid hits New York, the Metropolitan Museum may survive in a digital copy.

To make a digital image of the Unicorn Tapestries was one of the most difficult assignments that Bridgers had ever had. She put together a team to do it, bringing in two consultants, Scott Geffert and Howard

"The Unicorn in Captivity," *South Netherlandish, ca. 1495–1505. Wool warp, wool, silk, silver, and gilt wefts; 12 ft. 1 in. × 99 in. (368 cm. × 251.5 cm.). The Metropolitan Museum of Art, gift of John D. Rockefeller, Jr., 1937 (37.80.6).*

Goldstein, and two of the Met's photographers, Joseph Coscia, Jr., and Oi-Cheong Lee. They built a large metal scaffolding inside the wet lab and mounted on it a Leica digital camera, which looked down at the floor. The photographers were forbidden to touch the tapestries; Kathrin Colburn and her team laid each one down, underneath the scaffold, on a plastic sheet. Then the photographers began shooting. The camera had a narrow view; it could photograph only one three-by-three-foot section of tapestry at a time. The photographers took overlapping pictures, moving the camera on skateboard wheels on the scaffolding. Each photograph was a tile that would be used to make a complete, seamless mosaic of each tapestry.

Joe Coscia said that his experience with the Unicorn Tapestries was incomparable: "It was really quiet, and I was often alone with a tapestry. I really got a sense that, for a short while, the tapestry belonged to me." For his part, Oi-Cheong Lee felt his sense of time dissolve. "The time we spent with the tapestries was nothing—only a moment in the life of the tapestries," he said. It took them two weeks to photograph the tapestries. When the job was done, every thread in every tile was clear, and the individual twisted strands that made up individual threads were often visible, too. The data for the digital images, which consisted entirely of numbers, filled more than two hundred CDs. With other, smaller works of art, Bridgers and her team had been able to load digital tiles into a computer's hard drives and memory, and then manipulate them into a complete mosaic—into a seamless image—using Adobe Photoshop software. But with the tapestries that simply wouldn't work. When they tried to assemble the tiles, they found that the files were too large and too complex to manage. "We had to lower the resolution of the images in order to fit them into the computers we had, and it degraded the images so much that we just didn't think it was worth doing," Bridgers said. Finally, they gave up. Bridgers stored the CDs on a shelf and filed the project away as an unsolved problem.

One day in the spring of 2003, the distinguished mathematician and number theorist David Chudnovsky and his wife, the United Na-

Joseph Coscia, Jr., photographing one of the Unicorn Tapestries in the wet lab. Antonio Ratti Textile Center, The Metropolitan Museum of Art.
Image © Metropolitan Museum of Art

tions diplomat Nicole Lannegrace, were having lunch at the Bedford Hills estate of Errol Rudman, a hedge-fund manager and a patron of the Metropolitan Museum, and his wife, Diana. Walter Liedtke, the curator of European paintings at the Met, was there with his wife, Nancy, who is a math teacher. David Chudnovsky began talking about digital imagery. Walter Liedtke, who is a Rembrandt scholar, felt a little out of his depth. "I had the illusion that I actually understood what David was saying," he said. "But this was pearls before swine." Liedtke decided to put David Chudnovsky in touch with the Met's photographers. Not long afterward, David, along with Tom Morgan, a PhD candidate who was working with David and David's younger brother, Gregory Chudnovsky, visited Barbara Bridgers in the Met's photography studio. Bridgers told them, "I have a real-world problem for you."

YEARS EARLIER, I had gotten to know Gregory and David Chudnovsky. They were number theorists—they investigated the properties of numbers—and they designed and worked with supercomputers. The Chudnovsky brothers insisted that they were functionally one mathematician who happened to occupy two human bodies. I had become familiar with what American mathematicians referred to as the "Chudnovsky Problem"—the fact that, despite their stature and accomplishments, they couldn't seem to get permanent jobs in the academic world. The Chudnovsky Problem had been partially solved, and the Chudnovsky Mathematician was working at the Institute for Mathematics and Advanced Supercomputing, or IMAS, which was operating out of a laboratory room at Polytechnic University, in downtown Brooklyn. IMAS was essentially the Chudnovskys.

Gregory Chudnovsky was now in his early fifties, a frail man with longish hair and a beard that were going gray, and sensitive, flickering brown eyes. His health had continued to be uncertain. He had myasthenia gravis, a condition that he had developed in his teenage years and that kept him in bed or in a wheelchair much of the time. David was five years older than Gregory, a genial man, somewhat on the portly side, with a cultivated manner, and he had curly graying hair and pale blue eyes that could take on a look of sadness.

At the time I first got to know the Chudnovsky brothers, they had built a powerful supercomputer out of mail-order parts. It filled the living room of Gregory's apartment at the time, on 120th Street in Morningside Heights, near Columbia University. They said the machine didn't really have a name, but they told me I could refer to it as "m zero," in order to have something to call it. Gregory was then living in the apartment with his wife, Christine, who was an attorney at a midtown firm, and his mother, Malka Benjaminovna Chudnovsky, who was in declining health. Malka passed away in 2001. The Chudnovsky brothers were using their homemade supercomputer to calculate the number pi, or π, to beyond two billion decimal places. Pi is the ratio of the circumference of a circle to its diameter, and is one of the most mysterious numbers in mathematics. Expressed in digits, pi begins 3.14159 . . . and runs on to an infinity of digits that never repeat. Though pi has been known for more than three thousand years, mathematicians have been unable to learn much about it. The digits show no predictable order or pattern. The Chudnovskys had been hoping, very faintly, that their supercomputer might see one. However, the pattern in pi may be too complex and subtle for the human mind to grasp or for any supercomputer to find. In any event, the supercomputer used a lot of electricity. In the summer, it heated Gregory's apartment to above a hundred degrees Fahrenheit, so the brothers installed twenty-six fans around it to cool it down. The building superintendent had no idea that the brothers were investigating pi in Gregory's apartment.

While this was going on, neither of the brothers had a permanent academic job. They were untenured senior research scientists at Columbia and were getting along on grants and consulting fees, and their wives were also contributing to the family income. Their employment problem was complex: they were a pair, yet they would need to fit into a math department as a single faculty member. In addition, they were using computers, an activity that some mathematicians regarded as unclean. And Gregory was unable to live anywhere except in a room where the air is purified with HEPA filters. (He suffered from allergies that could prove life-threatening.) He would require special care and arrangements from a math department, and it wasn't clear how much teaching he'd be able to do.

One day the Chudnovskys were approached by a man named Jeffrey H. Lynford, who was the CEO of Wellsford Real Properties, a real estate investment firm. He and his wife, Tondra, had become fascinated with the Chudnovsky Problem and had become determined to try to solve it somehow—that is, to find jobs for the brothers. Jeff Lynford proposed trying to raise money to endow a chair of mathematics for the Chudnovskys at a university somewhere. In the end, after several years of fund-raising efforts, Jeff and Tondra Lynford gave $400,000 to Polytechnic University. This gift, along with others, including two gifts from a Dallas businessman named Morton H. Meyerson and a gift from the Wall Street investor John Mulheren, ended up being enough to partially endow the Institute for Mathematics and Advanced Supercomputing. (John Mulheren died in 2003.) Having an institute put the brothers on a more stable footing. Gregory and Christine moved to a specially modified apartment that had filtered air, in Forest Hills, and in 1999 they had a daughter, Marian.

At IMAS, the brothers set about building a new series of computers of Chudnovskian design. The latest of them was a powerful machine of a type called a cluster of nodes. It was similar in design to their original machine, m zero, but was very much more powerful. (The most powerful supercomputer today is tomorrow's desktop machine.) The brothers ordered the parts for their new "supercomputer" through the mail. It sat inside a framework made of metal closet racks and white plastic plumbing pipes, and the structure was covered with window screens—those parts of the machine came from Home Depot. The brothers referred to their computer cluster modestly as "nothing." Alternatively, they called it "the Home Depot thing." "To be honest, we really call it It," Gregory explained. "This is because It doesn't exactly have a name." They became interested in using It or the Home Depot thing to crack problems that had proved difficult, such as assembling large DNA sequences or making high-resolution 3-D images of works of art.

To try to make a perfect digital image of the Unicorn Tapestries sounded like an interesting problem to David and Gregory Chudnovsky. One day, David paid a visit to the Met. He left the Met carrying seventy CDs of the Unicorn Tapestries. He and Gregory planned to feed the data into the Home Depot thing and try to join the tiles together into seamless images of the tapestries. The images would be the largest and most complex digital photographs of any artwork ever made. "This will be easy," David said to Barbara Bridgers as he left. He was wrong.

"WE THOUGHT TO OURSELVES that it would be just a bit of number crunching," Gregory said.

But, David said, "it wasn't trivial."

The brothers had a fairly easy time setting up the tiles on the Home Depot thing. When they tried to fit the puzzle together, however, they found that the pieces wouldn't join properly; the warp and weft threads didn't run smoothly from one tile to the next. The differences were vast. It was as if a tapestry had not been the same object from one moment to the next as it was being photographed. Sutures were visible. The result was a sort of Frankenstein version of the Unicorn Tapestries. The Chudnovskys had no idea why.

David, in exasperation, called up Barbara Bridgers. "Somebody has been fooling around with these numbers," he told her.

"I don't think so, David. Nobody around here could do that."

David informed her that he and Gregory would need to obtain the complete set of raw data from the Leica camera. The next day, he went to the museum and collected, from Bridgers, two large blue Metropolitan Museum shopping bags stuffed with more than two hundred CDs, containing every number the Leica had collected from the Unicorn Tapestries. There were at least a hundred billion numbers in the shopping bags.

David took the subway back to Brooklyn, stopping off at a supermarket to buy some fruit. In the lab, he put his things down, and Gregory began going through them. "Where are the rest of the CDs?" he asked. One of the Metropolitan Museum bags was missing.

"My God! I left it on the subway," David said. Half the Unicorn Tapestries could have been anywhere between the Upper Bronx and Far Rockaway.

They began frantically calling the subway's lost and found. "Naturally, there was no answer," Gregory recalled. David retraced his route. He found the Met bag sitting under the lettuce bin at the supermarket. Apart from being slightly misted, the CDs were okay.

Then the brothers really began to dig into the numbers. Working with Tom Morgan, they created something called a vector field, and they used it to analyze the inconsistencies in the images.

The tapestries, they realized, had changed shape as they were lying on the floor and being photographed. They had been hanging vertically for centuries; when they were placed on the floor, the warp threads relaxed. The tapestries began to breathe, expanding, contracting, shifting. It was as if, when the conservators removed the backing, the tapestries woke up. The threads twisted and rotated restlessly. Tiny changes in temperature and humidity in the room caused the tapestries to shrink or expand from hour to hour, from minute to minute. The gold- and silver-wrapped threads changed shape at different speeds and in different ways from the wool and silk threads.

"We found out that a tapestry is a three-dimensional structure," Gregory went on. "It's made from interlocked loops of wool."

"The loops move and change," David said.

"The tapestry is like water," Gregory said. "Water has no permanent shape."

The photographers had placed a thin sheet of gray paper below the edge of the part of the tapestry they were shooting. Each time they moved the camera, they also moved the sheet of paper. Though the paper was smooth and thin, it tugged the tapestry slightly as it moved, creating ripples. It stretched the weft threads and rotated the warp threads—it resonated through the tapestry. All this made the tiles impossible to join without the use of higher mathematics and the Home Depot thing.

A color digital photograph is composed of pixels. A pixel is the smallest picture element that contains color. The Unicorn Tapestries are themselves made up of the medieval equivalent of pixels—a single crossing of warp and weft is the smallest unit of color in the image. The woven pixels were maddening because they moved constantly. The brothers understood, at last, that it would be necessary to perform vast seas of calculations upon each individual pixel in order to make a complete image of a tapestry. Each pixel had to be calculated in its relationship to every other nearby pixel, a mathematical problem known as an N-problem, big enough to practically choke It. They decided to concentrate on just one of the tapestries, "The Unicorn in Captivity." Gregory said, "This was a math problem similar to the analysis of DNA or speech recognition—"

"Look, my dear fellow, it was a real nightmare," David said.

Two of the tiles on the front of "The Unicorn in Captivity" had an eerie green tinge. While the photographers were shooting them, someone had opened a door leading to the next room, where a fluorescent light was on, causing a subtle flare. The Chudnovskys corrected the lighting by using the color on the back threads as a reference. "It took us three months of computation," Gregory said. "We should have just dropped it."

The final assembly of the image took twenty-four hours inside the nodes of It, the Home Depot thing. Gregory and David stayed up all night and ran It from their respective apartments. In the preceding months, each pixel in "The Unicorn in Captivity" had been crunched through many billions of calculations. That last night, there were billions more calculations. By sunrise, the machine had recaptured "The Unicorn in Captivity" in its entirety. The image was flawless.

ONE DAY IN AUTUMN, my wife and our three children and I went to Brooklyn and paid a visit to the Chudnovskys at IMAS, which is in Rogers Hall, on the Polytechnic campus. David met us in the lobby. He wore a starched white shirt, dark slacks, and Hush Puppies. We were joined by Tom Morgan, a quiet man in his fifties with blue eyes, gold-rimmed spectacles, and a ponytail. He handed us disposable booties, of the kind worn by medical people in operating rooms. Then we went in.

The IMAS lab was a large, loftlike industrial room, with computer-controlled shades and lights, and filtered air. The lights were dim. The walls were concrete and painted white. The brothers projected images on the walls, and they also used the walls as a whiteboard to perform calculations with erasable markers. The walls were covered with scribbles—work in progress. Most of the floor consisted of a vast digital image, in color, showing 115 different equations arranged in a vast spiral that breaks up into waves near the walls—a whirlpool of mathematics.

The equations were a type known as a hypergeometric series. Among other things, they rapidly produce the digits of pi. The Chudnovskys discovered most of them; others were found by the great Indian mathematician Srinivasa Ramanujan, in the early twentieth century, and by Leonhard Euler, in the eighteenth century. On one corner of the floor there was a huge digital image of Albrecht Dürer's engraving *Melencolia I*. In it, Melancholy is sitting lost in thought surrounded by various strange objects, including a magic square and a polyhedron with eight sides, called Dürer's solid. The Chudnovskys suspected that Dürer's solid is more curious mathematically than meets the eye.

Gregory Chudnovsky was half lying on the couch in his stocking feet, his body extended, facing the figure of Melancholy. His shoes, which were tucked inside surgical booties, had been left on the floor. He wore jeans and a soft leather jacket, and he seemed relaxed. Christine and Marian, who was five, were there. Marian was chattering and running around the lab happily. The effect of the child circling over her father's swirling equations was slightly vertiginous.

"At first, we were going to cover the entire floor with *Melencolia*, but it made people dizzy," Gregory said. "It made us dizzy, too. So we shrank it and moved it near the couch."

Close to the windows stood the cluster of bare computers, sitting

David and Gregory Chudnovsky in their laboratory at the Institute for Mathematics and Advanced Supercomputing (IMAS). The shadow profile behind them is that of Tom Morgan, their collaborator.

Dudley Reed

inside the frame of plumbing pipes and covered with window screens—It. There was a sound of many small whirring fans running inside It, keeping It cool. (I associate this sound with any room professionally occupied by the Chudnovskys.)

My daughter Marguerite, who was fifteen at the time, wanted to know which of the many equations in the floor was the one that the brothers had used to calculate pi with their previous supercomputer.

"Walk this way," David said to her. "Now you are standing on the equation."

She looked down. The equation swooped for a yard under her feet.

Because the Chudnovsky equation for pi is the most powerful and accurate formula for pi that's ever been found, it is also the most nearly perfect representation of pi known to humanity, other than a symbol. Whether some hidden order exists in pi is still something unknown to humanity.

At the far end of the room hung two thirteen-foot-tall sheets of cloth, mounted at right angles to each other, that displayed perfect digital images of, respectively, the front and back of "The Unicorn in Captivity." We walked up to the two pictures of the unicorn. First I looked at the front. I could see each thread clearly. The unicorn is spattered with droplets of red liquid that seems to be blood, although it may be pomegranate juice dripping from fruit in the tree. The threads in the droplets of blood are so deftly woven that they create an illusion that the blood is semitransparent. The white coat of the unicorn shines through.

Then I turned to the back of the tapestry. Here the droplets were a more intense red, with clearer highlights, and they seemed to jump out at the eye. The leaves of the flowers were a vibrant, plantlike green. (There are as many as twenty species of flowers in this tapestry. They are depicted with great scientific accuracy—greater than in any of the botany textbooks of the time. They include English bluebells, oxlip, bistort, cuckoopint, and Madonna lily. Botanists haven't been able to identify a few; it's possible that they are flowers that have gone extinct since 1500.) On the front, in contrast, the yellow dye in the green leaves has faded a bit, leaving them looking slightly bluish gray.

Gregory got up from the couch. David warned him to be careful, and he put his arm around Gregory's waist, while Gregory leaned on David and put his arm over David's shoulders. Then the Chudnovsky Mathematician moved slowly across the floor, until the brothers were standing (rather precariously) beside It. David explained that their image of the tapestry was a first step toward making even finer digital images of works of art. He said, "It's simple to take a picture of a Vermeer, but what you really want is an image of the painting in 3-D, with a resolution better than fifty microns"—that's about half the thickness of a human hair. "Then you can see the brushstrokes," he went on, raising his voice over the whirring of the fans inside It. "You can catalog the brushstrokes in the sequence they occurred, as they were laid down on top of one another."

When mathematicians work, they engage in intensely serious play.

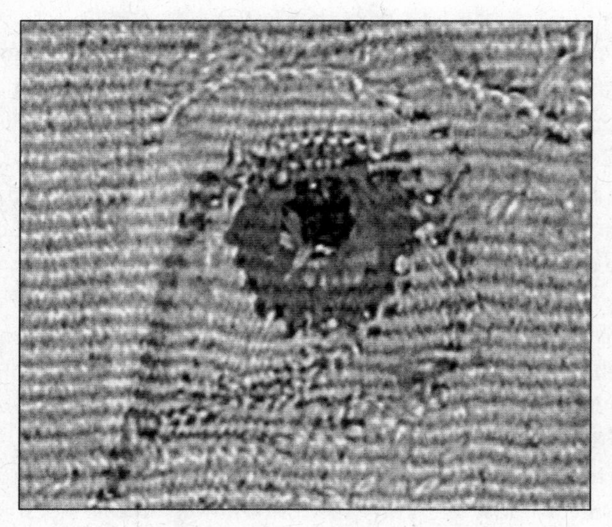

The eye of the unicorn, in reverse, on the back side of the tapestry. Detail from "The Unicorn in Captivity," South Netherlandish, ca. 1495–1505. Wool warp, wool, silk, silver, and gilt wefts; 12 ft. 1 in. × 99 in. (368 cm. × 251.5 cm.). The Metropolitan Museum of Art, gift of John D. Rockefeller, Jr., 1937 (37.80.6).

They follow their curiosity into problems that interest them. After playing with the unicorn, the Chudnovskys moved on.

"What are you doing now?" I asked.

David told me that they were working with IBM to design what may be the world's most powerful supercomputer. The machine, code-named C64, was being built for a United States government agency.

Among their projects, the Chudnovskys seemed to be involved with something called dark mathematics. Dark mathematics is classified mathematics. It's mathematics with national security implications that is done for government agencies. It isn't published in journals. Dark mathematics involves things like codes.

The superpowerful Chudnovskian C64 was rather like It, multiplied many times over, though nothing in C64 would come from Home Depot. When the government machine was finished, it would contain two million processors and fourteen thousand hard drives. It would use two and a half million watts of electricity—enough to power a few thousand homes. Two thousand gallons of water per minute would flow through the core of C64 to keep it cool. If the pumps failed, it would melt down in less than ten seconds.

ONE DAY, I went to see the Unicorn Tapestries in the physical universe, as distinct from the universe of numbers. It was a quiet winter afternoon at the Cloisters. The gallery where the tapestries hang was almost deserted. When I looked at them, each flower and plant, each animal, each human face took on a character of its own. The tapestries were alive with color and detail, full of velvety pools and shimmering surfaces. In the fence that surrounds the captive unicorn, tarnished silver, mixed with gold, gleamed in the grain of the wood. In comparison, the digital images, good and accurate as they were, had seemed flat. They had not captured the translucent landscape of the Unicorn Tapestries, as the weft threads dive around the warp, or the way they seemed to open into a world beyond the walls of the room.

Timothy Husband, the curator of the Cloisters, walked in. He was a tall, polished man in his late fifties, and had been at the Cloisters for thirty-five years. We sat down in one of the window seats facing the tapestries. "There is a luminosity and depth in them," he said quietly. "It didn't come about by chance on the part of the weavers."

I asked Husband how he felt when he was alone with the tapestries.

"That happens on Mondays, when the Cloisters is closed," he said. He spends anywhere from a minute to an hour with the tapestries. "It can be an exceedingly frustrating experience. One ponders so many questions about the tapestries for which there are no more answers today than there were when I was in graduate school." In some of the scenes, the unicorn may represent Christ. Alive and chained to the tree, after its apparent death in the hunt, it may speak of the immortality of the soul. Or the drops of blood may represent the pains of love. The truth is that the modern world has lost touch with the meanings in the Unicorn Tapestries. "Sometimes I come in here and try to pretend I have never read anything about them, never heard anything about them, and I just try to look at them," Husband said. "But it's not easy to shed that baggage, is it? And my other reaction, sometimes, is just to say, 'To hell with it, someday someone will figure them out.' And then there is a solace in their beauty, and one can stare at them in pure amazement."

The Self-Cannibals

One day in September 1962, a woman who here will be called Deborah Morlen showed up at the pediatric emergency room of the Johns Hopkins Hospital, in Baltimore, carrying her four-and-a-half-year-old son, Matthew. He was spastic and couldn't walk or sit up. As an infant, he had been diagnosed with cerebral palsy and developmental retardation. The hospital's pediatric emergency room was in the Harriet Lane Home for Invalid Children, an old brick building that stood in the center of the Johns Hopkins complex. Deborah Morlen sat down on a wooden bench in the waiting room. Her son lay stretched across her lap. He couldn't hold his head up, and his arms and legs thrashed around. His eyes were bright and restless. He was wearing mittens, though it was a hot day. Deborah Morlen had fastened the mittens tightly around his wrists with string, to keep them from falling off.

Eventually, a resident named Nancy Esterly saw Matthew in an examination booth, where she asked Deborah Morlen what was wrong.

Mattey was putting out strange-colored urine, Morlen told Esterly. "And there's, like, sand in his diaper," she said.

Nan Esterly removed the boy's diaper. It was stained a deep, bright orange, with a pink tinge. She touched the cloth and felt grit. She had no idea what this was, except that the pink looked like blood.

She began asking questions of Deborah Morlen, getting the boy's history. She eventually learned that Matthew had an older brother, Harold, who was also spastic and retarded, and also had had orange sand in his diapers. Harold was living at the Rosewood State Hospital, an institution for disabled children, outside Baltimore, while Matthew was living at home.

Since both brothers seemed to have the same condition, Nan Esterly thought it was likely that they had a genetic disease. A genetic disease is an inherited condition that runs in families, passing from parents to their offspring in their DNA. The human DNA, or human genome, is stored in the chromosomes, small elongated bodies in the nucleus of cells. Human cells contain two sets of twenty-three chromosomes (for a total of forty-six chromosomes in each human cell). The human genome contains a bit more than three billion letters of genetic code—enough letters to fill roughly a thousand complete editions of Edward Gibbon's *The Decline and Fall of the Roman Empire*. A gene is a stretch of DNA that holds the recipe for making a protein or group of proteins in the body.

In 1962, not a whole lot was known about genes and the human DNA. Even so, by that time Johns Hopkins pediatricians had begun discovering many previously unrecognized genetic diseases. Some of the doctors, perhaps not very kindly, would occasionally refer to children with genetic diseases as FLKs, or funny-looking kids. The story around Johns Hopkins was that every time a kid with an unusual appearance showed up in the emergency room, a new genetic disease would be found. Nan Esterly noticed that Matthew Morlen was wearing mittens, even though it was a warm day. She admitted the little boy to the hospital.

ESTERLY TOOK a sample of the boy's urine, and both she and an intern looked at it under a microscope. They saw that it was filled with crystals. They were beautiful—the crystals were clear as glass, and they looked like bundles of needles or like fireworks going off. They were sharp, and it was obvious that they were tearing up the boy's urinary tract, causing bleeding. Esterly and the intern pored over photographs of various kinds of crystals in a medical textbook, trying to identify them by their shape. The intern asked if the crystals might be uric acid, a waste product excreted by the kidneys; however, cystine, an amino acid that can form kidney stones, seemed the more likely candidate.

Esterly needed a confirmation of the diagnosis, so she carried the test tube upstairs to the top floor of the Harriet Lane Home, where William L. Nyhan, a pediatrician and research scientist, had a laboratory. "Bill Nyhan was the guru of metabolism," Esterly told me.

Nyhan, who was then in his thirties, had built some equipment that he was using to identify amino acids. He had been identifying amino acids in cancer cells while trying to find ways to cure cancer in children. "It was one of my impossible projects in cancer research," he told me. Nyhan later became a professor of pediatrics at the U.C. San Diego School of Medicine. "I love working with kids, but dealing with pediatric cancer was depressing, saddening, and, in truth, maddening," he said. Nyhan ran some tests on Matthew's urine, using the equipment he had designed. The crystals weren't cystine or any sort of amino acid. They proved to be uric acid.

A high concentration of uric acid in a person's blood can lead to gout, a painful disease in which crystals of uric acid grow in the joints and extremities, particularly in the big toe. Gout has been known since the time of Hippocrates, when ancient doctors recognized that it occurs mainly in older men. Yet the patient here was a little boy. Nyhan had a medical student named Michael Lesch working in his lab, and together they went downstairs to have a look at the boy with "gout."

Matthew lay in a bed in an open ward on the second floor of the Harriet Lane Home. The ward was filled with beds, and most of them were occupied by sick children. Matthew was a spot of energy in the ward, a bright-eyed child with a body that seemed out of control. The staff had tied his arms and legs to the bed frame with strips of gauze, to keep him from thrashing, and they had wrapped his hands in many layers of gauze. They looked like white clubs. Nurses hovered around the boy. "He knew I was a doctor and he knew where he was. He was alert," Nyhan recalled. Matthew greeted Nyhan and Lesch in a friendly way, but his speech was almost unintelligible: he had dysarthria, an inability to control the muscles that make speech. They noticed scarring and fresh cuts around his mouth.

They inspected Matthew's feet. No sign of gout. Then the boy's arms and legs were freed, and Lesch and Nyhan saw a complex pattern of stiff and involuntary movements, a condition called dystonia. Nyhan had the gauze unwrapped from the boy's hands.

Matthew looked frightened. He asked Nyhan to stop, and then he began crying. When the last layer was removed, they saw that the tips of several of the boy's fingers were missing. Matthew started screaming, and thrust his hands toward his mouth. With a sense of shock, Nyhan realized that the boy had bitten off parts of his fingers. He also seemed to have bitten off parts of his lips.

"The kid really blew my mind," Nyhan said. "The minute I saw him, I knew that this was a syndrome, and that somehow all of these things we were seeing were related."

Lesch and Nyhan began to make regular visits to the ward. Sometimes Matthew would reach out and snatch Nyhan's eyeglasses and throw them across the room. He had a powerful throw, apparently perfectly controlled, and it seemed malicious. "Sorry! I'm sorry!" Matthew would call, as Nyhan went to fetch his glasses.

The doctors persuaded Deborah Morlen to bring her older son to the hospital. Harold, it turned out, had bitten his fingers even more severely than Matthew and had chewed off his lower lip down into his chin, at the limit of the reach of his upper teeth. Both boys were terrified of their hands and screamed for help even as they bit them. The Morlen brothers, the doctors found, had several times more uric acid in their blood than normal children do.

Nyhan and Lesch visited the Morlen home, a row house in a working-class neighborhood in East Baltimore, where Matthew had been living with his mother and grandmother. "He was a well-accepted member of his little household, and they were very casual about his condition," Nyhan said. The women had devised a contraption to keep him from biting his hands, a padded broomstick that they placed across his shoulders, and they tied his arms to it like a scarecrow. The family called it "the stringlyjack." Matthew often asked to wear it.

Nyhan and Lesch also discovered that they liked the Morlen brothers. Lesch, who became the chairman of the department of medicine at St. Luke's–Roosevelt Hospital, in New York City, said, "Michael and Harold were really engaging kids. I really enjoyed being around them. I got beat up once by Matthew." He had leaned over the boy and asked him how he was feeling, and Matthew had slugged him in the nose. Lesch had staggered backward holding his nose while Matthew said, "Sorry! I'm sorry!"

TWO YEARS AFTER meeting Matthew Morlen, Nyhan and Lesch published the first paper describing the disease, which came to be called the Lesch-Nyhan syndrome. Almost immediately, doctors began sending patients they suspected of having the disease to Nyhan. Very few doctors had ever seen a person with Lesch-Nyhan syndrome, and boys with the disease were, and are, frequently misdiagnosed as having cerebral palsy. (Girls virtually never get it.) Nyhan himself found a number of Lesch-Nyhan boys while visiting state institutions for developmentally disabled people. When I asked him how long it took him to diagnose a case, he said, "Seconds." He went on, "You walk into a big room, and you're looking at a sea of blank faces. All of a sudden you notice this kid staring at you. He's highly aware of you. He relates readily to strangers. He's usually off in a corner, where he's the pet of the nurses. And you see the injuries around his lips."

WILLIAM NYHAN was eighty-one, a tall, fit-looking man with sandy-gray hair and blue eyes. He ran marathons until he was about seventy, half marathons after that; he was now the top-seeded tennis champion in his age class in southern California. He had a laboratory overlooking a wild canyon near the U.C. San Diego Medical Center. One day when I visited him, the Santa Ana wind was blowing in from the desert, and the air had an edgy feel. Two red-tailed hawks were soaring over the canyon, tracing circles in the air. The distinct movements of the hawks revealed a pattern of flight engraved in the birds' genetic code.

In the years since he had identified Lesch-Nyhan, William Nyhan had discovered or codiscovered a number of other inherited metabolic diseases, and he had developed effective treatments for some of them. He had figured out how to essentially cure a rare genetic disorder called multiple carboxylase deficiency, which could kill babies within hours of birth, by administering small doses of biotin, a B vitamin. Lesch-Nyhan, however, had proved to be more intractable.

Decades after the discovery of Lesch-Nyhan syndrome, it is still mysterious. It is perhaps the clearest example of a change in the human DNA that leads to a striking, comprehensive change in behavior. In 1971, William Nyhan coined the term "behavioral phenotype" to describe the nature of diseases like Lesch-Nyhan syndrome. A phenotype is an outward trait, or a collection of outward traits, that arises from a gene or genes—for example, brown eyes. Someone who has a behavioral phenotype shows a pattern of characteristic actions that can be linked to the genetic code. Lesch-Nyhan syndrome seems to be a window onto the deepest parts of the human mind, offering glimpses of the genetic code operating on thought and personality.

H. A. JINNAH, a neurologist at Johns Hopkins Hospital, has been studying Lesch-Nyhan syndrome for more than fifteen years. "This is a very horrible disease, and a very complex brain problem," he said to me one day in his office. "It is also one of the best models we have for trying to trace the action of one gene on complex human behavior."

A child born with Lesch-Nyhan syndrome seems normal at first, but by the age of three months has become a so-called floppy baby, and can't hold up his head or sit up. His diapers may have orange sand in them, and his body begins a pattern of writhing. When the boy cuts his first teeth, he starts using them to bite himself, especially at night, and he screams in terror and pain during these bouts of self-mutilation. "I get calls in the middle of the night from parents, saying, 'My kid's chewing himself to bits, what do I do?' " Nyhan said. The boy ends up in a wheelchair, because he can't learn to walk. As he grows older, his self-injurious behaviors become subtle and more elaborate, more devious. He seems to be possessed by a demon that forever seeks new ways to hurt him. He spits, strikes, and curses at people he likes the most—one way to tell if a Lesch-Nyhan patient doesn't care for you is if he's being very nice. (He wishes you would go away, so the Lesch-Nyhan part of him tries to keep you near him.) He eats foods he can't stand; he vomits on himself; he says yes when he means no. This is self-sabotage.

A few hundred boys and men alive in the United States today have been diagnosed as having Lesch-Nyhan syndrome. "I think I know most of them," Nyhan said. A boy known as J.J. whom Nyhan found in a state institution, where he'd been considered spastic and mentally retarded, ended up living in Nyhan's research unit for a year. He was a lively, gregarious child whose hands seemed to hate him with a demonic precision. Over time, his fingers had gotten into his mouth and nose, and had broken out and removed the bones of his upper palate and parts of his sinuses, leaving a cavern in his face. He had also bitten off several fingers. J.J. seemed happy most of the time, except when he was injuring himself.

J.J. died in his late teens; in the past many Lesch-Nyhan patients died in childhood or their teens, often from kidney failure. (Both Morlen brothers died young.) Nowadays they may live into their thirties and forties, but they often die from infections like pneumonia. Occasionally, a man with the disease flings his head backward with such force that his neck is broken, and he dies almost instantly.

A Lesch-Nyhan person may be fine for hours or days, until suddenly his hands jump into his mouth with the suddeness of a cobra strike, and he cries for help. People with Lesch-Nyhan feel pain as acutely as anyone else does, and they are horrified by the idea of their fingers or lips being severed. They feel as if their hands and mouths don't belong to them, and are under the control of something else. Some Lesch-Nyhan people have bitten off their tongues, and some have a record of self-enucleation—they have pulled out an eyeball or stabbed their eye with a sharp object such as a knife or a needle. (The eye is a soft target for the hand of a Lesch-Nyhan person.) When the Lesch-Nyhan demon is dozing, they enjoy being around people, they like being at the center of attention, and they make friends easily. "They really are great people, and I think that's part of the disease, too," Nyhan said. Some Lesch-Nyhan people are cognitively impaired and others are clearly bright, but their intelligence can't be measured easily. "How do you measure someone's intelligence if, when you put a book in front of him, he has an irresistable urge to tear out the pages?" Nyhan said.

. . .

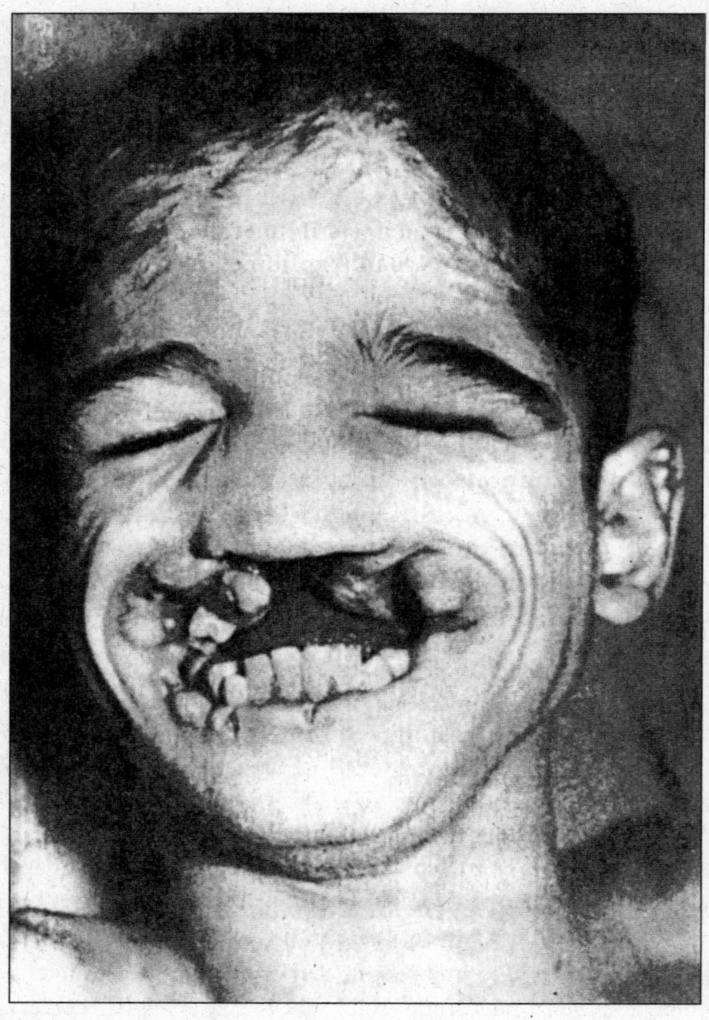

J.J., one of the earliest Lesch-Nyhan patients diagnosed by William L. Nyhan. His fingers pulled out the bones of his upper palate and sinuses, while his teeth removed several fingers. The bumps at the top of his ear were caused by gout; they are deposits of uric acid crystals known as tophi. J.J. was an outgoing child, very popular with his caregivers. He may be smiling in this picture, and he seems to have closed his eyes in anticipation of the camera flash.

William L. Nyhan

IN 1967, J. EDWIN SEEGMILLER, a scientist at the National Institutes of Health, and two colleagues discovered that in Lesch-Nyhan patients a protein called hypoxanthine-guanine phosphoribosyl transferase, or HPRT, which is present in all normal cells, doesn't seem to work. The enzyme is especially concentrated in deep areas of the brain, around the brain stem. The job of this enzyme is to help recycle DNA. Cells are constantly breaking DNA down into its four basic building blocks (represented by the letters A, T, C, and G, for adenine, thymine, cytosine, and guanine). This process produces compounds called purines, which can be used to form new code. When HPRT is absent or doesn't work, purines build up in a person's cells, where they are eventually broken down into uric acid, which saturates the blood and crystallizes in the urine.

In the early 1980s, two groups of researchers, one led by Douglas J. Jolly and Theodore Friedmann, decoded the sequence of letters in the human gene that contains the instructions for making HPRT. It includes 657 letters that code for the protein. Researchers also began sequencing this gene in people who had Lesch-Nyhan. Each had a mutation in the gene, but, remarkably, nearly everyone had a different one; there was no single mutation that caused Lesch-Nyhan. The mutations had apparently appeared spontaneously in each affected family. And in the majority of cases, the defect consisted of just one misspelling in the code. For example, an American boy known as D.G. had a single G replaced by an A—one out of over three billion letters of code in the human genome. As a result, he was tearing himself apart.

The HPRT gene is found on the X chromosome, which is the female sex chromosome and carries information that makes the person female. Women have two X chromosomes in each cell, and men have an XY pair. Lesch-Nyhan is an X-linked recessive disorder. This means that if a bad HPRT gene on one X chromosome is paired with a normal gene on the other X chromosome, the disease will not develop. A woman who has the Lesch-Nyhan mutation but carries it on only one of her X chromosomes doesn't develop the syndrome. Any son she has, however, will have a 50 percent chance of inheriting the syndrome (if he gets the bad X chromosome, he will have the disease; if he gets the good one, he won't), and any daughter will have a 50 percent chance of being a carrier. (Examples of this type of X-linked recessive disease include hemophilia and a form of red-green color blindness. Queen Victoria was a carrier of the hemophilia gene, but she didn't have hemophilia. Some of her male descendents had it.)

Other genetic mutations have been associated with profound behavioral changes. Rett syndrome, which affects mostly girls, is caused by a mutation in a gene that codes for the MeCP2 protein. People with the syndrome compulsively wring their hands and rub them together as if they were washing them. Children with Williams syndrome have an elfin appearance, an affinity for music and language, an extreme sensitivity for sound, and are very sociable. Williams syndrome is caused by a deletion of a bit of code from chromosome 7.

There is still great uncertainty about how much of a role genes play in major, common conditions such as depression or bipolar disorder. One wonders where obsessive-compulsive disorders come from, or such behaviors as compulsive hand-washing, compulsive neatness. Do some people suffer from OCDs that are caused by misspellings in their code? What about borderline personality disorder? How many mental illnesses are the result of errors in the code or certain combinations of errors? No one knows. It seems quite evident that a lot of human behavior is affected or governed by the blueprint of a person's DNA. Even where there is evidence of a family history of disease, scientists are unsure how a single gene could choreograph a suite of behaviors. There are roughly twenty-five thousand active genes in the human genome, each with about a thousand to fifteen hundred letters of code. The human genome could be thought of as a kind of piano with twenty-five thousand keys. In some cases, a few keys may be out of tune, which can cause the music to sound wrong. In other cases, if one key goes dead the music turns into a cacophony or the whole piano self-destructs.

The havoc that the Lesch-Nyhan mutation causes cannot easily be undone. Early on, Nyhan tried giving his Lesch-Nyhan patients allopurinol, a drug that inhibits the production of uric acid. The drug is effective with gout. It lowered the concentration of uric acid in Lesch-Nyhan patients, but it didn't reduce their self-injurious actions. The uric acid, it seemed, was another symptom, not a cause of the behavior.

Nyhan experimented with simple treatments, such as soft restraints, which seemed to relax patients and made them feel safer from themselves. Matthew Morlen had frequently asked to be tied into his stringlyjack. Nyhan also began recommending that Lesch-Nyhan children have their upper teeth removed, so that they couldn't bite off their lips and tongues as easily. "I'm profligate with those upper teeth," he said. This led to arguments with dentists. Some dentists would refuse to extract healthy teeth, even when the Lesch-Nyhan syndrome was explained to them.

I was visiting Nyhan in his office, and there was a lull in the conversation. He sat back in his chair and looked at me. I couldn't tell what he was thinking. Outside the window, the hawks were still riding thermals over the canyon. We had come to the end of knowledge about Lesch-Nyhan. The disease remained as mysterious and frightening as it had seemed on the day when Nyhan and Lesch had first seen it, nearly forty years earlier. But I had never seen it. "I can't imagine what it's like to have Lesch-Nyhan," I finally said to him.

"You could ask someone who has it," he replied.

I MET JAMES ELROD and Jim Murphy one winter day not long afterward. They were living next to each other in rented bungalows in a somewhat marginal neighborhood in Santa Cruz, California. James Elrod was then in his early forties, and Jim Murphy was just over thirty. They were close friends, except when their Lesch-Nyhan symptoms annoyed each other. (Murphy died in 2004; Elrod, who is fifty as I write this, became one of the oldest living people with Lesch-Nyhan.) The men were clients of Mainstream Support, a private company contracted by the state of California to help people with developmental disabilities live in community settings. Before he came to to Santa Cruz, James Elrod lived for eighteen years in a state institution in San Jose called the Agnews Developmental Center. Jim Murphy had spent all of his adult life in a California state institution in Sonoma. Mainstream employees, called direct-care staff, stayed with Elrod and Murphy around the clock, to help them with daily tasks and to make sure they didn't harm themselves. Elrod and Murphy had the authority to hire and fire their assistants and direct their work, though an assistant could refuse an order if he thought it would put the client in danger.

At the time, Mainstream was run by two business partners, Andy Pereira and Steve Glenn. "James and Jimmy are real down and gritty guys," Pereira said. "They are not sweet types. They're into fast cars and women." Steve Glenn confessed that he still had difficulty seeing into the labyrinth of Lesch-Nyhan. "There are these Lesch-Nyhan moments when you feel like you've kind of got it," he said. "James and Jimmy are pretty good at telling you when they think they're in danger of hurting themselves, but whenever they're doing something, you always have to ask, Is this James or Jimmy, or is it Lesch-Nyhan?"

James Elrod had a square, good-looking face, which was marked with scars, and he had brown, hyperalert eyes. His shoulders and arms were large and powerful, but the rest of his body seemed slightly diminished. One day at the Agnews Developmental Center, before he was with Mainstream, an attendant left him alone at a dining table for a few minutes. To his horror, his left hand picked up a fork and used it to stab his nose and gouge it out, removing most of his nose and permanently mutilating his face. "My left side is my devil side," he told me. When I met him, he wore black leather motorcycle gloves that had been reinforced with Kevlar. If he thought his left hand was threatening him or someone else, he would grab it or swat it with his right hand. He owned a pickup truck, and his assistants drove him around in it. He had a job working at a recycling facility. He also used to sell flowers on the Santa Cruz pier. He carried business cards explaining that he had a rare disease that compelled him to hurt himself. "I have injured myself in many ways including my nose, as you can see," the card said. "I will even try to hurt myself by getting into trouble with others." One day, a man bought some flowers from Elrod and said, "God bless you." "Eat shit," Elrod replied, and handed the man his business card. While crossing a street in his wheelchair, Elrod had been known to suddenly roll himself straight into oncoming traffic, yelling, "Slow down, you morons! Don't you know it's Lesch-Nyhan?" His assistants wrestled him to safety.

Elrod was sitting in front of his house in his wheelchair when I arrived. It was a sunny day. He offered me his right hand to shake. When

I gripped his glove, the right index finger collapsed. "You broke my finger!" he gasped. Then he grinned and explained that he didn't have that finger, as he had bitten it off some time ago.

I started laughing, but then regretted it. "I'm sorry to laugh," I said, imagining what he had done to himself.

No worries. He had given me a test, and I had passed it: I had laughed. "A lot of people get uptight when I do that," he said. "Kids love it. They want to break my finger again." We chatted for a while. "Hey, Richard—danger," he said.

"What's wrong?"

He was staring at my notes. I had been taking notes, as usual, in a little notebook. He cautiously pointed his finger at the mechanical pencil I was using. "Hey, Richard. Your pencil is scaring me." It had a sharp metal tip. "My hand could grab it and put it in my eye. Please step away from me and put your pencil down. Just listen."

I backed away from him, putting my notes and pencil in my pocket. But then he said, "You'd better go see my neighbor. He's waiting for you."

It only occurred to me later that James Elrod might have entangled me in an act of self-sabotage. He had been looking forward to meeting a writer and describing his disease. He had been waiting in his driveway for me. Because he wanted very much to tell me about his disease, the Lesch-Nyhan part of him had threatened to grab my pencil and puncture his eye with it, had ordered me to put down my notes, and finally had sent me away to interview another man with Lesch-Nyhan syndrome, instead of himself. Was I reading too much into it? It was hard to say. It seemed as if he might be playing a chess game with himself in which he was doing everything he could to put his desires in checkmate.

JIM MURPHY was sitting in his wheelchair at a table in the living room of his house. He had been waiting for me, too. An assistant named Michael Roth was cutting up pancakes and feeding them to him with a spoon. Murphy was a bony man with dark hair and a lean, handsome face. He had a neatly trimmed goatee and a crew cut, and his eyes were mobile and sensitive-looking. His lips were missing. Two of his brothers had also had Lesch-Nyhan; they had died when they were young. "Jimmy will be shy when you first meet him," one of his sisters told me on the phone. I could expect to hear a lot of swearing, though. "He doesn't mean it," she said. "When he swears at me, I just say, 'I love you, too.' "

That day in Santa Cruz, Murphy stared at me out of the corners of his eyes, with his head involuntarily thrown back and turned away, braced against a headboard. His hands were stuffed into many pairs of white socks, and he wore soft, lace-up wrestler's shoes. His chest heaved against a rubber strap that held him in place. He started throwing punches at me, and he kicked at me. He seemed to be enduring his disease like a man riding a wild horse. The wheelchair shook.

I kept back. "It's nice to meet you," I said.

"Fuck you. Nice to meet you." Jim Murphy had a fuzzy but pleasant-sounding voice. His speech was very hard to understand. He looked at Michael Roth. "I'm nervous."

"Do you want your restraints?" Roth asked.

"Yeah."

Roth placed Murphy's wrists in soft cuffs fastened with Velcro, and he placed his legs in cuffs, as well. The wheelchair trembled and rattled as his limbs fought against the cuffs.

"I'm a little nervous, too," I said, sitting down on the couch.

"I don't care. Good-bye."

I stood up to leave.

Roth explained, however, that this was one of those Lesch-Nyhan situations where words mean their opposite.

I sat down again. "Do you want me to call you Jim or Jimmy?" I asked.

His answer was blurry.

"I'm having a little trouble understanding you," I said.

"I duhcuh . . ."

"You don't care?"

He repeated his words several times until he saw that I understood. He was saying: "I don't care between Jim or Jimmy. Either's fine."

"I sort of like Jim better, myself."

He said something I couldn't understand.

"What's that?"

He repeated his words. He was saying, "Do you want me to call you Richard or Dick?"

"Oh, I don't care. Either is fine," I said.

An impolite grin spread across Murphy's face. "I'm going to call you Dickhead. That's your new dickname."

I burst out laughing.

Meanwhile Roth, the assistant, seemed not to be hearing a word of our conversation.

Later, Jim Murphy explained what his disease was like. "You try to tick everybody off, and then you feel bad when you do it," he said. Slowly I became better able to identify his words. "If you get too close to me, I could—" He said something indecipherable.

"I'm sorry, what?"

"Coldcock you, Richard. I'll say, 'Get my water,' and I'll give you a sucker punch."

A pair of red boxing gloves was hanging on the wall. Every day, his assistants placed him on a wrestling mat on the floor, where he rolled around and did stretches and then boxed with them. "I could definitely whip you," he told me.

Later, Jim Murphy asked if I would like to go for a walk with him around Santa Cruz, alone, without his assistant. I said sure. Then he asked me to take off his restraints. "Don't worry, Richard," he said.

Feeling nervous about the situation, I opened his cuffs. His arms flew out and started waving around, but he didn't throw any punches at me. I pushed him out the door. We went down a driveway and came to a cul-de-sac, where we had to make a decision, to turn either left or right.

"Go right," he said. I started to turn him to the right "No! Left," he said. So I turned left. "No! No! Right!"

We were trapped in a Lesch-Nyhan hall of mirrors, filled with reflections of desire and repulsion. "Which way do you really want to go, Jim?"

"Left."

"Are you totally sure?"

"Left! Left!"

The leftward path led through a gate. As we passed through the gate, one of his sock-covered hands shot out and struck the gate, hard. He had compulsively hurt himself. I apologized to him and said, "I guess I really should have gone the other way."

"Not your fault."

I began pushing him along in the street, keeping away from mailboxes. (I was afraid he'd try to hit one.) I was beginning to be able to figure out his speech.

A young woman driving a vintage Ford Mustang convertible passed us.

He waved to her and called out a greeting, and she waved back; they apparently knew each other.

Murphy seemed entranced. "Did you see that? She waved at me. She's beautiful," he said in a slushy voice. "She likes me. I love Mustangs," he added.

"I used to have a '65 Ford Falcon when I was in college," I said as I pushed him along. "It was a '65 Mustang under the skin. It had the same engine and interior as the Mustang, but the body had this weird shape."

"Yeah!" He grinned. "I love those old Falcons. Do you still have it?"

"I gave it to one of my brothers."

"Does he like it?"

"Well, he sold it to some kid for fifty dollars."

"Aw, no!" Murphy said. "I bet you want it right now."

"I sure do."

We ended up at a corner grocery store, and while Murphy chatted with a woman at the cash register, who was a friend of his, I went to the cooler case to buy him a Coke. "Get whatever you want. It's my treat," he said. "Take my wallet." He used his eyes to indicate a pocket in the wheelchair where he kept his wallet.

"Thanks," I said.

"Fuck you. You're welcome."

Later, at James Elrod's house, I was sitting on his back porch and chatting with him, and he pointed out various plants he was cultivating in pots. The pots sat on shelves where he could reach them from his wheelchair. Elrod was a passionate gardener. He dug in the soil with his gloved hands; he didn't dare hold a tool of any sort. His assistants often did gardening tasks for him, while he told them what to do.

"Jim gave me a new nickname," I remarked to James Elrod.

"Yeah? What is it?"

"It's 'Dickhead,' " I said.

A glow of delight lit up Elrod's face. "Hey, guess what. I'm going to call you . . . it's . . ." He couldn't get the word out. "Jack—" he mumbled.

"My name is Jack?" I asked.

"Eee! Aww!"

"What?"

"It's Jackass, Dickhead!" He laughed uproariously.

JAMES ELROD was born in 1957 in a small town in northern California, where his father worked as a laborer in a rice-drying warehouse. As a child, he was never able to walk, and doctors diagnosed him with "cerebral palsy." He had an older brother, Robert, who also had "cerebral palsy," but, unlike James, Robert was considered to be mentally retarded. (Robert Elrod died in 1998.) James attended regular elementary school until he was in fifth grade, when his parents put him in a special education program.

"I couldn't walk, but I could scoot," James explained. Scooting meant crab-walking on his hands and feet with his stomach facing up and his rear end bumping along the ground. James and his older sister, who here will be called Marjorie, were very close as children. When they were children, Marjorie was James's steadfast companion. She pulled James around in a wagon. Marjorie towed James to school every morning in the wagon, and she towed him home in the afternoon. James and his sister have remained close.

Their father drank heavily. He would come home drunk and become enraged with James. "My dad used to hit me with a belt on my bare back," he said. "I'll forgive him for it. But he never forgave me for being what I am."

The family went camping in the Sierra Nevada, and James fished with his father and helped him hunt deer; he learned how to dress a deer. He hauled wood for the campfire, scooting around on the ground while balancing pieces of wood on his stomach. He was well-liked pretty much wherever he went.

His grandparents lived nearby in the same town, in a farmhouse near the railroad tracks. James was fond of his grandparents and spent much of his free time visiting with them. He liked to work in his grandmother's garden with her, scooting up and down the rows of tomatoes, pulling weeds and helping out. Hoboes drifted by on the railroad tracks, walking along the back side of James's grandparents' property. James's grandmother would hire the hoboes to pull weeds, and her payment was a sandwich. James pulled weeds along with the hoboes. He also delivered milk and cookies to the hoboes, perched on his stomach, and he hung out by the railroad tracks for hours, talking with the hoboes. Eventually, James's grandmother invited a drifter named Herbie to come stay with the family and work for them, in exchange for room and board. For the rest of his life, Herbie made his home in a gardening shed behind James's grandparents' house.

Things were not so smooth at home, where his father became in-

James Elrod.

Christopher Reeves

creasingly violent. "My father was drinking a lot," James recalled. "He forced booze down my throat and then he lunged at my mother." This terrified and enraged the boy, but he was helpless, and all he could do was scuttle around on the floor; he couldn't protect his mother from the violence. Until then, he had not been known to engage in self-injury. "The first Lesch-Nyhan episode I remember was when I was about ten," he said. "My mother was taking me home from school in the wagon and I jumped out the wagon and tried to hurt myself. Afterward, the school nurse called up Child Protection Services and told them that I was getting hurt a lot when I was around my mother." These official suspicions of his mother tormented him as a child, because he knew very well that she wasn't hurting him, he was hurting himself.

He went to high school, attending special ed classes, but as a teenager he became harder for his parents to manage. They eventually made him a ward of the state of California and had him put in a rest home for elderly people, hoping that he would receive good care there. The staff of the nursing home began giving him medication. He was a young man, full of energy, who ended up on sedatives in a room with Alzheimer's patients. "There was only one other young person in the rest home besides me, and she was a sixteen-year-old girl," Elrod said. "The worst part of it was they wouldn't let me do any gardening." This seemed to disturb the demon, and one day he threw himself through a glass window, cutting himself badly. His parents had him transferred to a nursing home in Sacramento. There he was allowed to have a garden, and he gave the products of his garden to the residents of the home—they got vegetables to eat and flowers to brighten their rooms. "The garden had really good soil," he said. "I planted all kinds of bulbs there. I had tomatoes, cucumbers, squash, and beans growing in it, and good, big pumpkins. It had these really big sunflowers growing in a row."

His sister, Marjorie, visited him there and discovered that the staff was giving him heavy doses of drugs—overmedicating him, in her view. "I raised Cain every time I went to see him there," Marjorie said to me. "The staff said they were just doing what they had to do. They told me that I needed to have an appointment before I could see him.

I told them I was going to come see my brother whenever I wanted. That was when James starting smashing his face on the table." The nursing home couldn't handle him.

At twenty-one, Elrod ended up at the Agnews Developmental Center in San Jose. Gardening was out of the question at the state institution. Elrod ended up being paired with a roommate who was profoundly retarded and couldn't speak, though he made continual noise. "My roommate was a screamer, twenty-four/seven," he said.

"It was just horrible for him," Marjorie said.

In the state institution, his hands began to go out of control. They began attacking his face. He bit his hands in order to protect his face from attacks by them—he needed to hurt his hands in order to scare them away. He began to realize that his hands would stop at nothing if they got a chance. It was in the state institution one day when his left hand hacked off his nose with a fork. The Lesch-Nyhan demon came brilliantly awake, and turned its gaze on Elrod with murderous intent.

THERE HAVE BEEN about twenty autopsies of Lesch-Nyhan patients over the years. Their brains appeared to be perfectly normal. "It's a problem in the connections, in the way the brain functions," H. A. Jinnah, the Johns Hopkins neurologist, said. He had gotten interested in Lesch-Nyhan as a scientific enigma, but he quickly found that families all over the United States were sending Lesch-Nyhan patients to him, hoping for help. He couldn't turn them away, and he had ended up looking after a large number of boys and young men with Lesch-Nyhan who were his patients. "It's an orphan disease," he explained. "Almost nobody studies it."

During some of the autopsies, doctors had tested samples of brain tissue to see if they contained a normal balance of neurotransmitters—chemicals that are used for signaling between nerve cells. In the Lesch-Nyhan brains, a lemon-sized area containing structures called the basal ganglia, near the center of the brain, had 80 percent less dopamine—an important neurotransmitter—than a normal brain. The basal ganglia are wired into circuits that run all over the brain; they affect a wide range of behaviors: motor control, higher-level thinking, and eye movement, as well as impulse control and enthusiasm.

"People with Lesch-Nyhan have an excess number of involuntary movements," Jinnah said. "It's as if they are stepping on the gas too hard when they try to do something. If you ask them to look at a red ball, for instance, their eyes go to everything except the red ball, and they can't explain why. Then, if you introduce a yellow ball into their field of view but you don't say anything about it, they watch the yellow ball." The moment you draw their attention to it, however, they look away. Most Lesch-Nyhan people can't read, and even watching television may be difficult or impossible for them—their eyes refuse to follow the action on the television screen.

"Lesch-Nyhan is at the far end of a spectrum of self-injurious behavior," Jinnah went on. "You and I and everyone who is human, we all do things that are bad for us. We'll sit down in front of the television and eat a quart of ice cream. We all have self-injurious impulses, too. Driving a car, we can have a strange impulse to drive it the wrong way and smash it into something." People who are afraid of heights can experience the sensation that they may throw themselves from a high place, driven by some impulse they can't control. Edgar Allan Poe called such promptings "the imp of the perverse." The imp may be signals coming out of the basal ganglia of the brain. "Many people bite their fingernails," Jinnah said. "They'll tell you it's gross and that they don't want to do it—'Sometimes I get nervous and start biting my fingernails,' they'll say. There are people who chew their lips nervously. Now let's turn up the volume a little: some people bite their cuticles. Turn up the volume a little more: some people bite their cuticles until they bleed. Now let's turn the volume *way* up. Now you have someone biting off tissue and bone in his fingers, biting off the whole finger, and chewing his lips off. Where, in this spectrum of behavior, is free will?"

We can all think of things we've done to ourselves that make us cringe. It's the inexplicable choice that brings about the very thing we feared the most. A shadow of self-sabotage seems to move behind the scenes on the stage of history. Napoleon's invasion of Russia comes to mind. The mass behavior of much of the German population before and during the Second World War has a Lesch-Nyhanish quality. Israelis and Palestinians fight, neither side seeming able to compromise sufficiently to truly benefit their self-interest; Palestinian factions fight one another, like the hand attacking the face. Wars are about power, control of resources, and unrighted wrongs, to be sure, but they also seem to reveal a glimpse of self-mutilation in the human species. Perhaps there is a little bit of Lesch-Nyhan in each of us.

In a medical sense, Lesch-Nyhan looks like Parkinson's disease reversed. People with Parkinson's have trouble starting physical actions. Lesch-Nyhan people start actions too easily and can't stop an action once it starts. Because Parkinson's is also associated with a deficiency in the dopamine of the basal ganglia, scientists have looked to each disease for clues to the other.

In 1973, a researcher named George Breese, at the University of North Carolina School of Medicine, was working with rats that modeled Parkinson's disease. He was treating newborn rats with compounds that changed the dopamine levels in their brains when, to his surprise, the rats started chewing off their paws. He had inadvertently created a Lesch-Nyhan rat. He was absolutely horrified, yet felt he might be tantalizingly close to a cure for Lesch-Nyhan. "I'll not go further into the details of what the rats were doing. They weren't biting their mouth tissues the way human patients do," Breese told me. If he gave the self-injuring rats another compound, they stopped biting their paws—that is, he found a way to reverse the symptoms. "We treated the rat the moment we saw the animal make the first pinprick injury to its paws," he said. The compound, however, has never been approved for use on humans.

In April 2000, Takaomi Taira, a neurosurgeon at the Tokyo Women's Medical University, performed brain surgery on a nineteen-year-old man with Lesch-Nyhan. The young man was living with his parents in a district north of Tokyo. In addition to exhibiting self-injurious behavior, he had the spastic, stiff, thrashing movements of dystonia. "These dystonic movements were getting more severe almost by the day, and his parents were getting desperate," Taira said to me recently. He decided to perform a procedure called deep-brain stimulation to try to calm down these movements.

Deep-brain stimulation was developed by doctors more than twenty years ago for treating people with Parkinson's disease. One or more thin wires are carefully navigated through the brain until they stop in a part of the basal ganglia called the globus pallidus (the "pale globe"). The wires are connected to a battery pack, which is implanted under the skin of the patient's chest, and a faint, pulsed current of electricity runs through them into the globus pallidus, numbing a spot the size of a pea. The patient feels nothing. The procedure often helps calm the tremors in Parkinson's patients' hands and limbs, and helps them walk more easily.

"After the surgery, the boy's dystonic movement completely disappeared," Taira said. He sent him home with the deep-brain stimulator, feeling that the operation had helped. Several months later, the young man's parents told Taira that he had stopped biting himself. He was still in a wheelchair, and his uric-acid levels remained high, but he was reading comic books and watching television, and seemed to be enjoying life as never before. "It was completely unexpected, remarkable, almost unbelievable," Taira said. A few years later, the young man suddenly began biting his hands again, and the parents brought him back. "I checked the device and found that the battery was flat. I replaced the battery, and his symptoms were controlled again," Taira said.

A group at the Research Group on Movement Disorders in Montpellier, France, led by a neurosurgeon, Philippe Coubes, has given deep-brain-stimulation implants to five Lesch-Nyhan patients. His method involves the insertion of four wires into the brain. "So far, we have three patients who are doing very well and two who are having an intermediate response—the response of one of those is not poor but is not as good as the others," Coubes said. "I'm not sure we will be able to control all their behaviors over the long term, but we are in the process of getting a better understanding of deep-brain stimulation for these patients." The imp of the perverse could be put to sleep, but nobody knew how to make it go away.

Scientists aren't sure why deep-brain stimulation seems to work in some patients. Indeed, the results are a reminder of how obscure the workings of the brain still are. William Nyhan was cautious about the procedure's potential. "I see these kids as fragile, and they don't respond very well to surgical invasions," he said.

At Johns Hopkins, though, Jinnah was anxious to begin a study on a group of Lesch-Nyhan patients using deep-brain stimulation. He still needed to get funding and to receive approval from the federal government. (The procedure has not been specifically approved for Lesch-Nyhan patients.)

Jinnah has never had an easy time getting funding and attention for Lesch-Nyhan research. "People ask me, 'Why not study more common diseases?' My answer is that if we neurologists did that we'd all be studying Alzheimer's disease, Parkinson's disease, and strokes. There are thousands of other brain diseases out there, and they're all orphans. But these rare diseases may teach us something new about the brain, something relevant to the common brain diseases that affect so many people."

I WENT BACK several times to visit James Elrod and Jim Murphy and began helping their staff with daily tasks. Elrod spat in my face a few times, and gave me a left jab to the jaw that made me see stars, for which he apologized afterward. Once, his Kevlar-covered fingers closed on my skin like pliers; he apologized while we both worked to get them loose. Murphy, at his thirty-first birthday party, planted his face in his birthday cake, then punched me in the groin so hard that I collapsed to the floor. Nevertheless, I came to like them a lot.

Murphy had a record of making trouble in shopping malls. Malls put him in a bad mood, especially around Christmastime. "Too many people around. They make me nervous," he explained to me.

One time, his assistants took him to a mall to do some Christmas shopping. A man dressed as Santa Claus was sitting in a snow scene that day, with children lining up to meet him. Murphy told his assistants that he would like to have his picture taken with Santa (one of them had a camera). They didn't see how they could refuse the client's request. They parked Murphy's wheelchair in the line of children, and Murphy cautioned the children to watch out for his arms and legs. (Neither Murphy nor Elrod had been known to lash out at a child.)

Murphy got to the head of the line. The Santa asked Murphy if he'd like to sit on his lap.

Murphy said yes. The assistants placed him on the Santa's lap. The assistant with the camera, a young man named Dan Densley, got ready to take a picture.

"Ho, ho, ho! What do you want for Christmas?" Santa asked.

"A woman," Murphy answered, and delivered a punch to Santa's jaw. Santa's beard seemed to explode, and his eyeglasses went flying. The assistants grabbed Murphy and rolled him out of the mall at a dead run.

JIM MURPHY had a passion for off-road driving, which he was not usually able to indulge. One day, I showed up in Santa Cruz in a rented Ford Expedition with four-wheel drive. A woman named Tracye Overby came along as Murphy's assistant, while another assistant, Christopher Reeves, accompanied Elrod. Into the back of the vehicle we put a cooler holding a roast chicken and beer, and I drove the group to a dried-out lake bed near Watsonville called College Lake, which we'd heard was a good place for four-wheeling. On the way, I stopped to get directions from a California state trooper. "I would not advise going there with disabled people," he said.

College Lake turned out to be a mile-wide expanse of clay covered with sand. The lake bed had a dark, wet-looking center and was surrounded by thickets of willows. I edged the Expedition out onto the sand, while Murphy and Elrod began to egg me on from the backseat. "Go faster," Murphy said. "You're driving like an old lady." I gunned the engine, the Expedition leaped forward, and we raced across the sand. When I turned sharply, the men shouted with delight. I started cutting cookies and performed a figure eight, then aimed the vehicle toward the center of the lake and floored the accelerator. With the engine roaring, we passed the sunken carcass of a truck, buried up to its roof in clay. The Expedition began slowing down, even though the engine was running at full power. Then the vehicle began tipping over—which was when I realized that we were driving across quicksand. If we stopped, we would go down. I floored it again and turned back for the shore, but it was too late. We slowed to a halt with the engine making an extreme wailing sound and the wheels spinning. The Expedition settled down until its axles were buried; then the engine suddenly died. There was a moment of complete silence. Then the men erupted with profanities. "Jackass!" Elrod yelled.

The two helpers seemed unperturbed. "This is just the nature of our work," Chris Reeves said. "Everything that you plan never goes as you planned it."

I opened the door and tested my weight on the sand. At least the Expedition seemed to be floating on it. I got out my cell phone and started calling local towing services.

Elrod leaned over me. "Tell them you've got two Lesch-Nyhan guys in the car, and we're going to cut your head off."

After a number of calls, it became clear that the local towing companies wouldn't touch College Lake. "I'll just have to get somebody to tow *me* out," one man said.

"We could call my lawyer," Elrod suggested.

"Forget it!" Jim Murphy said. "Your lawyer won't do us any good. Call the fire department. Help! Fire! Fire!"

I tried an outfit called Speed of Light Towing. A young guy answered the phone and said he couldn't do it.

I told him it was a desperate situation. "I've got two disabled guys with me. They're in wheelchairs and can't walk. We're in the middle of the lake," I said.

"What?" he said. He needed to hear that again. "All right. I'll try," he finally said. I would have to pay him when he arrived, no results guaranteed.

"I'm nervous," Murphy said. Blood dribbled out of his mouth—he was biting himself. Tracye Overby lifted him out of the car, carried him across the sand to the shade of some willows, and sat down, holding him in her lap. She wiped the blood from his mouth with a napkin, cradled his head in her arms, and began singing a song to him:

Oh Mama, oh Papa, I'm feeling so down,
I can't believe there's no milk in this town . . .
I'll do anything for a glass of white milk.
I live in a town, a town without milk.

"I'm sorry, Tracye," Murphy said.

"What are *you* apologizing for?" She patted his stomach.

Murphy looked at me. "It's your fault, Richard. No more off-roading with you." He began to laugh.

"What's so funny?" Overby said to him.

"I've had enough, Tracye. I'm leaving. I'm getting out of here right now. Good-bye." He thrashed around on his back, kicking and laughing. He couldn't even sit up.

Elrod, sitting in the front seat of the Expedition, began laughing, too. Their laughter drifted through the silence of College Lake. The men were connoisseurs of what I had done: I had ignored the advice

Stranded in College Lake.
Tracye Overby (left) and Christopher Reeves (right) with Jim Murphy lying between them. He seems quite entertained. (Note his sock-wrapped hand resting on Reeves's left arm.)
Richard Preston

of a police officer and driven two disabled men at high speed into the mud. They saw something familiar in my behavior.

I popped a beer and handed it to Elrod. "Are you okay?"

"Just fine."

I opened a beer for myself. "There's chicken in the cooler. Do you want some?"

He ate pieces of chicken with his gloves on, and asked me not to hand him any pieces that had sharp bones.

It was a cloudless day in spring, without a breath of wind. The Santa Cruz Mountains stretched into the distance, blanketed with many colors of green. Canyons wandered down through the mountains, jammed with the dark spires of redwoods. A flock of coots burst from the willows and flew straight across the lake bed, heading west toward the sea. Overhead, violet-green swallows dodged and looped. The birds were behavioral phenotypes, their movements controlled by their genes. A streak of dust appeared in the east, at the edge of the lake, and extended toward us: something moving fast. Speed of Light Towing was coming for us. A battered pickup truck with fat tires stopped a good distance away. A young man got out. He walked over to us, dragging a chain and stomping his feet on the sand as he went along, testing the sand. His name, it turned out, was Robert, and he was the same person I'd spoken with. He glanced at the Lesch-Nyhan men. "Hey, how are you?" he said in friendly way.

"Fine," Murphy answered in a slushy voice.

Robert said to me quietly, "What's with them?"

"It would take a while to explain," I said.

"We'll need to dig," Robert added.

All of us (except Elrod and Murphy) got down on our knees and began scooping sand from under the vehicle with our hands, Robert working along with us. Ten minutes later, we'd dug a tunnel under the vehicle. Robert crawled into it and got the chain hooked into the chassis. He crawled out. "Start your vehicle," he said to me. "Give it gas when the chain goes tight. If the chain breaks, watch out. It could come back through the windshield and cut your head off."

He started his truck, gunned the engine. The truck shot forward in a running start, and chain snapped tight.

There was a sound like a gunshot; the chain had broken. It whiplashed back and hit the Speed of Light truck with a booming sound that echoed over the lake, leaving a dent in the tailgate.

Robert got out and studied the damage to his truck. He seemed philosophical about it. "I should've used the big one." Then he reached into the bed of his truck and unfurled a massive chain—the big one. He dragged it slowly over to us, and got it attached. "This'll definitely pull *something* out of here. I just hope it's not your axle." He floored his truck, the chain went tight, and the Expedition was yanked out,

Jim Murphy soon after being rescued from quicksand by Speed of Light Towing, Watsonville, California. He is sticking his tongue out at the photographer. "No more off-roading with you."

Richard Preston

bouncing and fishtailing over the lake bed. Perhaps having Lesch-Nyhan syndrome is like being stuck in mud all your life while waiting for help that never arrives.

A COUPLE OF YEARS LATER, Jim Murphy came down with pneumonia, and he went downhill fast. When it was clear that he was dying, I called him to say good-bye. As he got on the line, I could hear a hubbub of voices in the background. A lot of people had come to see him, and they were milling around in his hospital room and in the hallway. He seemed to be handling his situation calmly. "I'll be all right," he said to me, and added, "Take care with your driving."

Another day, before Jim Murphy died, I visited James Elrod. Tracye Overby, who was working as Elrod's assistant that day, needed to change the silk liners he wore inside his motorcycle gloves.

Elrod did not like to see his bare hands. He asked me to hold his wrists while Overby removed his gloves. It occurred to me that I had never seen his hands. The hands that emerged were pale, with spindly fingers that had been gnawed close to the bone in places, and a finger was missing. "Danger," he said. His eyes took on a strange, bright, blank look. He was staring at the right hand. His arm was tense and trembling. As if a magnet were pulling it, the hand moved toward his mouth. His mouth opened, wider and wider, baring his teeth. . . . "Help!" he called in a muffled voice.

We threw ourselves on Elrod. It took all our strength to restrain his hand. As soon as we got control of it, he relaxed. Overby got the gloves back on him.

"Nobody knows about this disease. Every day I'm hoping for a cure," Elrod said. "I wanted you to see that."

Glossary

arthropod A segmented invertebrate with a hard exoskeleton made of chitin. Examples: spiders, mites, insects, crustaceans.

basal ganglia An area in the lower part of the brain, the size of a lemon, that influences many aspects of behavior.

behavioral phenotype A pattern of actions and behavior traceable to the influence of a **gene** or genes in the DNA.

biohazard space suit A pressurized whole-body protective suit made of soft, flexible plastic, with a soft helmet, worn by researchers working in **Biosafety Level 4 (BL-4)** laboratories.

Biosafety Level 4 (BL-4) The highest level of biocontainment security.

Celera Genomics A division of the Applera Corporation devoted to genomic and proteomic discovery to advance the practice of medicine.

CDC Centers for Disease Control and Prevention, a federal organization, part of the Department of Health and Human Services, responsible for the detection and prevention of human disease, headquartered in Atlanta, Georgia.

chromosome A small elongated body in the nucleus of a cell in which a portion of the organism's DNA is tightly coiled, for storage. Human cells have two sets of twenty-three chromosomes (for a total of forty-six chromosomes).

Chudnovsky Mathematician, the The brothers David and Gregory Chudnovsky assert that functionally they are a single mathematician who happens to occupy two human bodies.

decon shower A chemical decontamination shower used in the air lock entry/exit module of a Biosafety Level 4 lab.

DNA sequencing Determination of the sequence of **nucleotides,** or letters, in a strand of DNA.

ecotone A boundary-like habitat in nature where different kinds of ecosystems come into contact and mix.

Eddington number, the The number of protons and electrons in the observable universe. The Eddington number is roughly 10^{79}, or a 1 followed by seventy-nine zeros; it was first proposed by the British physicist Sir Arthur Eddington in 1938.

epistaxis Nosebleed.

EST Expressed sequence tag. An easily identifiable sequence of letters in DNA, typically located near the end of a **gene.**

gene A stretch of the DNA, typically a thousand to fifteen hundred letters long, that holds the recipe for making a protein or a group of proteins in an organism.

genetic disease An inherited illness or impairment that is passed from parents to their offspring in a gene or genes in the DNA. A genetic disease is not contagious.

genome, human The total amount of DNA that is spooled into a set of chromosomes in the nucleus of every typical human cell.

genomics The sequencing and study of genes in DNA.

gout A disease, first identified by doctors at the time of Hippocrates, in which crystals of uric acid build up in the extremities, especially the big toe, causing severe pain.

Home Depot thing, the (or **It**) A powerful computer built by mathematicians David and Gregory Chudnovsky.

host An organism in or on which a **parasite** lives.

hot zone, hot suite A Biosafety Level 4 biocontainment laboratory.

HPRT protein A protein produced by all cells in the human body, used for recycling purines (by-products of the processing of DNA). When HPRT is absent from cells, due to a defect in a gene, the result in humans is **Lesch-Nyhan syndrome.** The full name of this protein is hypoxanthine-guanine phosphoribosyl transferase.

Human Genome Project, the A nonprofit international research consortium that deciphered the complete sequence of nucleotides, or letters, in the human DNA.

IMAS Institute for Mathematics and Advanced Supercomputing, at Polytechnic University, Brooklyn. Principally occupied by David and Gregory Chudnovsky (**the Chudnovsky Mathematician**).

Institute, the Nickname for the United States Army Medical Research Institute of Infectious Diseases (USAMRIID), at Fort Detrick, Maryland.

It *See* **Home Depot thing.**

J. Craig Venter Institute A nonprofit institute dedicated to research in genomics, founded and run by genomic scientist J. Craig Venter.

Lesch-Nyhan syndrome A rare genetic disease, almost invariably expressed in males, in which the patient engages in compulsive acts of self-injury. It was first characterized in 1962 by medical researcher William L. Nyhan and medical student Michael Lesch at Johns Hopkins Hospital in Baltimore.

Ludolphian number, the The same number as **pi** (**π**). The name is derived from Ludolph van Ceulen, a German mathematician of the seventeenth century who approximated pi to thirty-five decimal places and had the digits engraved on his tombstone.

m zero A powerful supercomputer constructed largely of mail-order parts by the mathematicians David and Gregory Chudnovsky. Predecessor to the **Home Depot thing.**

Mbwambala A patch of disturbed woodland about three miles long and half a mile wide that wanders along a stream about six miles southeast of the city of Kikwit, Democratic Republic of the Congo.

National Institutes of Health (NIH) A federally funded collection of medical research institutes situated on a campus in Bethesda, Maryland, that both conducts and funds many billions of dollars in medical research every year.

nucleotide An information-carrying building block, or "letter," of DNA. There are four nucleotides in DNA: adenine, thymine, cytosine, and guanine; they are designated by the letters A, T, C, and G.

number theory The mathematical study of the properties of numbers.

parasite An organism that lives on or inside another organism, its **host,** and feeds on the host, being harmful to the host or of no benefit to it.

pi (π) The ratio of the circumference of a circle to its diameter. Expressed in decimals, pi goes 3.14159 . . . and continues infinitely, without periodically repeating. Pi is a **transcendental number.**

pubic symphysis An area in the lower front of the pelvis where the pelvic bones join in a suture filled with cartilage.

red diarrhea, the The local Congolese term for an Ebola virus infection during the 1995 outbreak in Kikwit, Congo.

self-mutilation, compulsive Uncontrollable physical self-injury, such as self-biting. In **Lesch-Nyhan syndrome** it arises ultimately from a defect in the gene that codes for the **HPRT** protein, though the exact mechanism of the disease is unknown.

strebelid flies Parasitic wingless flies that crawl and live on bats. A conjectured possible natural host of the Ebola virus.

supercomputer One of the world's most powerful computers for its time.

TIGR The Institute for Genomic Research, a nonprofit research institute dedicated to sequencing **genomes,** now part of the **J. Craig Venter Institute** in Rockville, Maryland.

transcendental number A number that is not the exact solution to any polynomial equation that has a finite number of terms with integer coefficients. *See* **pi.**

tubular cast, throwing a Expelling through the anus a sleevelike lining of the intestines and rectum.

Unicorn Tapestries, the Seven tapestries of large size and exceptional preservation and beauty (though one of them is now in fragments), originally woven around 1500 in Brussels or Liège, now hanging in the Cloisters Museum in New York City. The

Unicorn Tapestries are considered to be among the great works of art of all time.

virus A disease-causing agent smaller than a bacterium consisting of a shell made of proteins and membranes and a core containing DNA or RNA. A virus is a parasite that can replicate only inside living cells, using the machinery of the cell to make more copies of itself.

warp, weft Strong, straight noncolored threads (warp threads) and delicate undulating colored threads (weft threads) are woven to form a tapestry. In many late medieval tapestries, including the **Unicorn Tapestries,** the warp threads run horizontally and the weft threads run vertically.

wet lab An underground room at the New York Metropolitan Museum of Art where tapestries and works of fabric art are washed, conserved, and photographed.

Zarate procedure A surgical procedure whereby the bones of the pelvis are cut in front, at the location of the **pubic symphysis,** the cut running through a suture of cartilage there. It causes the pelvis to spring open. The Zarate procedure is a crude but effective way of releasing a baby stuck in the birth canal.

Acknowledgments

The principal thanks in this book must be given to the people who are portrayed in it. They often patiently and generously submittted to the sort of tedious questioning that I gave Nancy Jaax when I examined her hands. I'm especially grateful to: Nancy Jaax; "Jeremy"; "Martha"; Gregory and Christine Chudnovsky; David Chudnovsky and Nicole Lannegrace; the late Malka Benjaminovna Chudnovsky; the late Herbert Robbins; Richard Askey; William T. Close; Will Blozan; Heidi Blozan; Rusty Rhea; Kristine Johnson; Tom Remaley; Tim Tigner; Lee Frelich; Carolyn Mahan; Richard Evans; James Åkerson; Christopher Asaro; Stephen C. Sillett; D. Scott Sillett; Robert Van Pelt; J. Craig Venter; Claire Fraser; Hamilton O. Smith; Marshall R. Peterson; James D. Watson; Eric S. Lander; Norton Zinder; Francis Collins; Gene Meyers; Jeffrey and Tondra Lynford; Morton H. Meyerson; Tom Morgan; Peter Barnet; Barbara Bridgers; Scott Geffert; Joseph Coscia, Jr.; Oi-Cheong Lee; Timothy Husband; William L. Nyhan; Michael Lesch; Nancy Esterly; James Elrod; James Elrod's sister; the late and beloved Jim Murphy; all the members of the Murphy family I met, who gave so generously of their time and thought; Andy Pereira; Steve Glenn; Tracye Overby; Michael Roth; Christopher Reeves; Brad Alerich; H. A. Jinnah; Takaomi Taira; Philippe Coubes.

Many thanks to Tim Bartlett, my editor at Random House, who is the overall editor of this book. Many thanks also to Tina Bennett and Lynn Nesbit at Janklow & Nesbit Associates. At *The New Yorker,* past and present, I'm grateful to the following people for their contributions to various parts of this book: Robert Gottlieb, Tina Brown,

David Remnick, John Bennet, Sharon DeLano, Dorothy Wickenden, Amy Davidson, Peter Canby, Martin Baron, Ann Goldstein, Elisabeth Biondi, Elizabeth Culbert, and the late Miss Eleanor Gould (Eleanor Packard); while the following checkers worked on certain parts: Hal Espen ("The Mountains of Pi"); Christopher Jennings and Michael Peed ("A Death in the Forest"); Bill Vourvoulias and Daniel Hurewitz ("The Search for Ebola"); Andy Young ("The Human Kabbalah"); Marina Harss ("The Lost Unicorn"); and Lila Byock and Jessica Rosenberg ("The Self-Cannibals"). Any errors of fact in this book are my responsibility, though where I got things right, very often a checker was involved.

My wife, Michelle, and our children, Marguerite, Laura, and Oliver, with their endless curiosity and openness to new things, inspired this book. They were also present for some of the interviews in "The Lost Unicorn," and they have had their own friendship with the Chudnovsky family and asked their own questions. Michelle, who worked as a checker at *The New Yorker*, inspired me in fact-checking. She continues to inspire me in far greater ways than that.

Read on for an excerpt from
The Wild Trees by Richard Preston

Nameless

ONE DAY IN THE MIDDLE of October 1987, a baby-blue Honda Civic with Alaska license plates, a battered relic of the seventies, sped along the Oregon Coast Highway, moving south on the headlands. Below the road, surf broke around sea stacks, filling the air with haze. The car turned in to a deserted parking lot near a beach and stopped.

A solid-looking young man got out from the driver's side. He had brown hair that was going prematurely gray, and he wore gold-rimmed spectacles, which gave him an intellectual look. His name was Marwood Harris, and he was a senior at Reed College, in Portland, majoring in English and history. He walked off to the side of the parking lot and unzipped his fly. There was a splashing sound.

Meanwhile, a thin, tall young man emerged from the passenger side of the car. He had a bony face, brown eyes, and a mop of sun-streaked brown hair, and he wore a pair of bird-watching binoculars around his neck. Scott Sillett was a junior at the University of Arizona, twenty-one years old, visiting Oregon during fall break. He took up his binoculars and began to study a flock of shorebirds running along the surf.

The interior of the Honda Civic was made of blue vinyl, and the back seat was piled with camping gear that pressed up against the windows. The pile of stuff moved and a leg emerged, followed by a curse, and a third young man struggled out and stood up. "Mardiddy, this car of yours is going to be the death of us all," he said to Marwood Harris. He was Stephen C. Sillett, the younger brother of Scott Sillett. Steve Sillett was nineteen and a junior at Reed College, majoring in biology. He was shorter and more muscular than his older brother. Steve Sillett had feathery light-brown hair, which hung out from under a sky-blue bandanna that he wore tied around his head like a cap. He had flaring shoulders, and his eyes were dark brown and watchful, and were set deep in a square face. The Sillett brothers stood shoulder to shoulder, looking at the birds. Their bodies were outlined against decks of autumn rollers coming in, giving off a continual roar. Scott handed the binoculars to his younger brother, and their hands touched for an instant. The Sillett brothers' hands had the same appearance—fine and sensitive-looking, with deft movements.

Scott turned to Marwood: "Marty, I think your car should be called the Blue Vinyl Crypt. That's what it will turn into if we fall off a cliff or get swiped by a logging truck."

"Dude, you're going to get us into a crash that will be biblical in its horror," Steve said to Marwood. "You need to let Scott drive." (Steve didn't know how to drive a car.)

Marwood didn't want Scott's help with the driving. "It's a very idiosyncratic car," he explained to the Sillett brothers. In theory, he fixed his car himself. In practice, he worried about it. Lately Marwood had noticed that the engine had begun to give off a clattering sound, like a sewing machine. He had also become aware of an ominous smell coming from under the hood, something that resembled the smell of an empty iron skillet left forgotten on a hot stove. As Marwood contemplated these phenomena and pondered their significance, he wondered if his car needed an oil change. He was fairly sure that the oil had been changed about two years ago, in Alaska, around the time the license plates had expired. The car had been driven twenty thousand miles since then, unregistered, uninsured, and unmaintained, strictly off the legal and mechanical grids. "I'm worried you'll screw it up," he said to Scott.

Steve handed the binoculars to his older brother and climbed into the back of the Blue Vinyl Crypt. "Dudes, let's go," he said. "We need to see some tall redwoods."

They planned to go backpacking in one of the small California state parks that contain patches of ancient coast redwood forest.

None of the young men had ever seen a redwood forest. Steve seemed keyed up.

THE COAST redwood tree is an evergreen conifer and a member of the cypress family. Its scientific name is *Sequoia sempervirens.* It is sometimes called the California redwood, but most often it is simply referred to as the redwood. No one knows exactly when or where the redwood entered the history of life on earth, though it is an ancient kind of tree, and has come down to our world as an inheritance out of deep time. A redwood has furrowed, fibrous bark, and a tall, straight trunk. It has soft, flat needles that become short and spiky near the top of the tree. The tree produces seeds but does not bear flowers. The seeds of a redwood are released from cones that are about the size of olives. The heartwood of the tree is a dark, shimmery red in color, like old claret. The wood has a lemony scent, and is extremely resistant to rot.

Redwoods grow in valleys and on mountains along the coast of California, mostly within ten miles of the sea. They reach enormous sizes in the mild, rainy climate of the northern stretches of the coast. Parts of the North Coast of California are covered with temperate rain forest. A rain forest is usually considered to be a forest that gets at least eighty inches of rain a year, and parts of the North Coast get more than that. A temperate rain forest has a cool, moist, even climate, not too hot or cold. Redwoods flourish in fog, but they don't like salt air. They tend to appear in valleys that are just out of sight of the sea. In their relationship with the sea, redwoods are like cats that long to be stroked but are shy to the touch. The natural range of the coast redwoods begins at a creek in Big Sur that flows down a mountain called Mount Mars. From there, the redwoods run up the California coast in a broken ribbon, continuing to just inside Oregon. Fourteen miles up the Oregon coast, in the valley of the Chetco River, the redwoods stop.

The coast redwood is the tallest species of tree on earth. The tallest redwoods today are between 350 and close to 380 feet in height—thirty-five to thirty-eight stories tall. The crown of a tree is its radiant array of limbs and branches, covered with leaves. The crown of a super-tall redwood has a towering, cloudy form, and the crowns of the tallest redwoods can sometimes look like the plume of exhaust from a rocket taking off.

Botanists make a distinction between the height of a tree and its overall size, which is measured by the amount of wood the tree has in its trunks and limbs. The largest redwoods, which are called redwood giants or redwood titans, are usually not the very tallest ones. In this way, they are rather like people. A football player is often bigger than a basketball player—more massive, that is. The basketball player is taller and more slender. So it is with redwoods. The tallest redwoods are often slender, and so they aren't the largest ones. Even so, the most massive redwoods (the redwood titans) are among the world's tallest trees anyway, and are more than thirty stories tall. Today, almost no trees of any species, anywhere, reach more than three hundred feet tall, except for redwoods. The main trunk of a redwood titan can be as much as thirty feet in diameter near its base.

Many people who are familiar with coast redwoods have seen them in the Muir Woods National Monument, in Marin County, just north of the Golden Gate Bridge. Muir Woods, which is visited by nearly a million people every year, is a tiny patch of virgin, primeval redwood forest, and it is like a small window that reveals a glimpse of the way much of Northern California looked in prehistoric times. Though the redwoods in Muir Woods are hauntingly beautiful trees, they are relatively small and are not very tall, at least for redwoods. The redwoods you can see in Muir Woods are nothing like the redwood titans that stand in the rain-forest valleys of the North Coast, closer to Oregon. These are the dreadnoughts of trees, the blue whales of the plant kingdom.

Nobody knows the ages of any of the living giant coast redwoods, because nobody has ever drilled into one of them in order to count its annual growth rings. Drilling into an old redwood would not reveal its age, anyway, because the oldest redwoods seem to be hollow; they don't have growth rings left in their centers to be counted. Botanists suspect that the oldest living redwoods may be somewhere between two thousand and three thousand years old—they seem to be roughly the age of the Parthenon.

THE ROAD became the California Coast Highway, and the Sillett brothers and Marwood Harris drove past Jedediah Smith Redwoods State Park, in Del Norte County. They didn't stop to look at the redwoods there. They went through Crescent City, a tired-looking town. They passed a Carl's Jr. fast-food restaurant, and a lumber mill, and bars and taverns, dark in daylight, where you could get a beer for a dollar and maybe get a fractured skull for nothing.

The redwood forests around Crescent City had been logged. The road went past stretches of open land covered with bare stumps, and past seas of young redwood trees growing on timber-company land, which looked like plantations of fuzzy Christmas trees. Here and there on the ridges were a few last stands of virgin, ancient redwoods, looming above everything else. They looked like Mohawk haircuts.

The road entered Del Norte Coast Redwoods State Park, and the highway was suddenly lined with extremely tall redwoods. Steve Sillett began thrashing around in the back of the Crypt. "Stop the car! I'm getting out."

Marwood pulled off to the side of the road. Steve squeezed out of the back seat and took off, running into the forest. Scott and Marwood waited in the car.

"What's he doing?"

"He's looking at the trees."

"Oh, God."

They rolled down the windows. "Steve! We're not there yet! Get back in the fricking car!"

TWENTY MILES farther down the road, they came to Prairie Creek Redwoods State Park. The park occupies a sliver of wrinkled terrain, eight miles long and four miles wide, lying along the Pacific Ocean on the northern edge of Humboldt County. The North Coast along those parts is covered with rain forests, and the forests are often hidden in clouds and fog. The beaches along the North Coast are made of gray sand, gnawed by waves the color of steel. The beaches rise into bluffs, which become the California Coast Ranges, a maze of ridges and steep, narrow valleys, clad with deep temperate rain forest. The forest is dominated by coast redwoods.

As they entered the park, Steve was hunched over, staring at a map. Marwood slowed to a crawl. Trucks whipped past them. Steve ordered Marwood to stop, and he pulled off the highway and rammed the Blue Vinyl Crypt into the underbrush, to get it out of sight. They were planning to camp in some wild spot among the redwoods, but it is illegal to camp in the redwood parks except in a few public campgrounds, and they didn't want the rangers to notice their car.

They put on their backpacks and hurried along a trail that went westward, climbing toward a ridge and the ocean, passing through a redwood forest. The trees had stony-gray bark. They looked like the columns of a ruined temple. The ground was made up of rotting redwood needles, and it was covered with sword ferns—tall, stiff ferns—growing chest high. Everywhere there were spatters of redwood sorrel—small, emerald-green plants with heart-shaped leaves.

The trail came to the crest of a ridge and dropped down into a valley that opened toward the ocean. As they went over the ridge, the sound of trucks on the highway faded away. A hush came over the world, and it grew dark. There was no sunlight at the bottom of the redwood forest, only a dim, gray-green glow, like the light at the bottom of the sea. The air grew sweet, and carried a tang of lemons. They became aware of a vast forest canopy spreading over their heads.

PHOTO: ROBERT LEWIS

RICHARD PRESTON is the bestselling author of *The Hot Zone*, *The Cobra Event*, *The Demon in the Freezer*, and *The Wild Trees*. A writer for *The New Yorker* since 1985, Preston won the American Institute of Physics Award and is the only non-doctor ever to have received the CDC's Champion of Prevention Award. He also has an asteroid named after him. He lives near New York City with his wife and three children.